Fault Location on Transmission and Distribution Lines

Fault Location on Transmission and Distribution Lines

Principles and Applications

Swagata Das
Surya Santoso
Sundaravaradan N. Ananthan

IEEE PRESS

WILEY

This edition first published 2022
© 2022 John Wiley & Sons Ltd

The right of Swagata Das, Surya Santoso, and Sundaravaradan N. Ananthan to be identified as the authors of this work has been asserted in accordance with law.

Registered Offices
John Wiley & Sons, Inc., 111 River Street, Hoboken, NJ 07030, USA
John Wiley & Sons, Ltd, The Atrium, Southern Gate, Chichester, West Sussex, PO19 8SQ, UK

Editorial Office
The Atrium, Southern Gate, Chichester, West Sussex, PO19 8SQ, UK

For details of our global editorial offices, customer services, and more information about Wiley products visit us at www.wiley.com.

Wiley also publishes its books in a variety of electronic formats and by print-on-demand. Some content that appears in standard print versions of this book may not be available in other formats.

Library of Congress Cataloging-in-Publication Data

Names: Das, Swagata, author. | Santoso, Surya, author. | Ananthan, Sundaravaradan N, author.
Title: Fault location on transmission and distribution lines : principles and applications / Swagata Das, Surya Santoso, Sundaravaradan N. Ananthan.
Description: Hoboken, NJ : Wiley, 2022. | Includes bibliographical references and index.
Identifiers: LCCN 2021030668 (print) | LCCN 2021030669 (ebook) | ISBN 9781119121466 (cloth) | ISBN 9781119121473 (adobe pdf) | ISBN 9781119121497 (epub)
Subjects: LCSH: Electric fault location. | Power transmission.
Classification: LCC TK3226 .D3173 2022 (print) | LCC TK3226 (ebook) | DDC 621.319/2–dc23
LC record available at https://lccn.loc.gov/2021030668
LC ebook record available at https://lccn.loc.gov/2021030669

Cover Design: Wiley
Cover Image: © REDPIXEL.PL/Shutterstock

Set in 9.5/12.5pt STIXTwoText by Straive, Chennai, India
Printed and bound by CPI Group (UK) Ltd, Croydon, CR0 4YY

C9781119121466_051121

Contents

Preface

The purpose of the power system is to is to deliver electrical power from generators to loads through transmission and distribution lines. Today, electrical power is a necessity of everyday life and is expected to be present whenever we need it. Unfortunately, this expected service may not always be available, particularly when there is a short-circuit fault on the power system. The outage time will equal the time taken by the utility to find the fault plus the time taken to make repairs.

Fault-locating algorithms can shorten the time taken to find faults on transmission and distribution lines. These lines cover long distances. Manually patrolling the line to locate the fault can be quite time consuming. Rough terrains and bad weather can make it that much harder to locate the fault. Knowing where to look for the fault saves significant time and expedite service repair and restoration.

Today, there are many fault-locating algorithms to choose from. While all algorithms have the same objective, which is to find the fault with the highest accuracy, each algorithm makes different assumptions and uses different data to arrive at this result. What is the best fault-locating algorithm? The correct answer is that it depends. It depends on the data available for fault location, the system to which the algorithm will be applied to, and the nature of the fault.

The purpose of this book is to explain how the different fault-locating algorithms work, identify what data they need, and discuss the sources of error. This information can help the reader make an informed decision about what is the best fault-locating algorithm for their system. Several real-world examples are presented that show how to apply these algorithms to fault event data and solve for the fault location. The book also covers the additional benefits of fault location that can help improve power system performance and reliability.

We hope that practicing engineers and students will find this book helpful. Please feel free to reach out to us if you have any comments for improvement or find any errors.

Finally, we would like to sincerely thank Electric Power Research Institute for giving us fault event data used in this book.

29 October 2021
San Antonio, Texas
Austin, Texas
Houston, Texas

Swagata Das
Surya Santoso
Sundaravaradan N. Ananthan

About the Companion Website

This book is accompanied by a companion website:

www.wiley.com/go/das/faultlocation

The website includes the following materials:

- Fault Location Suite in MATLAB

1

Introduction

The power system is a complex network of generators, transformers, transmission lines, distribution feeders, loads, and other electrical components. The purpose of this network is to deliver electrical power from generators to loads through transmission lines and distribution feeders. Today, electrical power is a necessity of everyday life and is expected to be present whenever we flip the light switch, charge our phones, and turn on other gadgets. Unfortunately, this expected service may not always be available, particularly when there is a fault on the power system.

Faults are abnormal conditions on the power system that disrupt the normal flow of electrical power from generators to loads. Lightning, animal contact, tree contact, and adverse weather such as strong winds and winter storms are some of the major reasons for power system faults. Utilities take many preventive steps such as installing shield wires and surge arresters to divert the energy of lightning strikes, putting up animal guards, and trimming trees at periodic intervals to minimize the chances of a fault. In spite of all these measures, faults are inevitable on the power system. So when they occur, all efforts must be made to locate the fault as quickly as possible, make repairs, and restore power. This is why fault location is so important and critical to improving power system reliability.

We begin this chapter by explaining the types of faults that can occur on the power system and their root cause. We then move our focus to fault-locating algorithms. We discuss their aim and importance, their principles, their implementation in the field, and their evaluation criteria. Finally, we discuss how to choose the best fault-locating algorithm from among the many algorithms that have been proposed in the literature.

1.1 Power System Faults

Faults are abnormal conditions on the power system that cause voltage, current, frequency, and power to deviate from their nominal values. Protective relays are typically used to detect and isolate these faults as quickly as possible to return the power system back to normal operating conditions. The Institute of Electrical and Electronics Engineers (IEEE) defines a protective relay as "a device whose function is to detect defective lines or apparatus or other power system conditions of an abnormal or dangerous nature and to initiate appropriate control action" [1]. Protective relays use current transformers (CTs) and potential

Fault Location on Transmission and Distribution Lines: Principles and Applications, First Edition.
Swagata Das, Surya Santoso, and Sundaravaradan N. Ananthan.
© 2022 John Wiley & Sons Ltd. Published 2022 by John Wiley & Sons Ltd.
Companion website: www.wiley.com/go/das/faultlocation

transformers (PTs) to monitor the state of the power system. When a fault is detected, they send a trip command to circuit breakers, which then open to isolate the fault. In low-voltage distribution systems, fuses are often used instead of protective relays and circuit breakers to detect and isolate faults that create an overcurrent condition. A fuse is defined by IEEE as "an overcurrent protective device with a circuit-opening fusible part that is heated and severed by the passage of the overcurrent through it."

Faults experienced by the power system can be of two types, series faults and shunt faults. Series faults usually occur when there is an open circuit on one or two phase conductors during load conditions. Because the open circuit occurs in series with the phase conductor, these faults are known as series faults. Series faults can be caused by broken jumpers or when all three poles of a circuit breaker pole are unable to close during a manual or an automatic close operation. They can also be caused by a blown fuse. For example, Fig. 1.1 shows a distribution transformer being protected by high-side fuses. If one or two high-side fuse blows due to an overcurrent condition, it will result in a series fault. During a series fault, the current in the faulted phase decreases due to loss in load while the healthy phase continues to carry load current. The voltage and frequency of the faulted phase also increase as compared to the healthy phase. While series faults do not result in high magnitude currents to flow in the faulted phases, they make the power system unbalanced, causing unbalanced currents to flow in the power system. The heat generated by the unbalanced currents can damage transformers and motors. References [2–4] explain how protective relays can be set up to detect and isolate series faults.

Shunt faults occur when there is a shunt connection between one or more phase conductors to the ground or between each other. The shunt connection creates a short-circuit condition allowing current to flow through an alternate, lower impedance path. The lower impedance causes the current in the faulted phase to dramatically increase while the voltage of the faulted phase decreases. Because shunt faults are more common and more damaging than series faults, this book will focus on locating shunt faults.

There are four types of shunt faults (see Fig. 1.2). Single line-to-ground faults (also referred to as single phase-to-ground faults) occur when one of the three phase conductors makes contact with the ground wire or the grounded piece of an equipment. Seventy to eighty percent of all faults are single line-to-ground faults, making this the most common fault type [5]. Line-to-line faults (also referred to as phase-to-phase faults) occur when two phase conductors make contact with each other. Double line-to-ground faults (also referred to as double phase-to-ground faults) occur when two phase conductors make contact with each other and the ground wire or the grounded piece of an equipment. Three-phase faults occur when all three phase conductors make contact with each other,

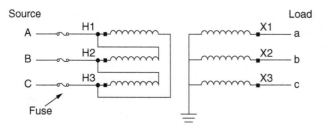

Figure 1.1 A distribution delta/wye-grounded transformer being protected by high-side fuses. A blown fuse will result in a series fault.

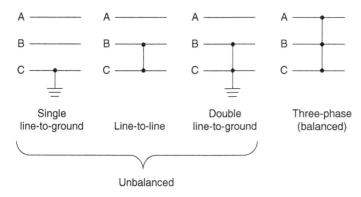

Figure 1.2 The four types of shunt faults.

with or without ground connection. This fault type is quite rare and is most often the result of human errors. Three-phase faults are referred to as balanced faults as the fault involves all three phases. The other fault types involve one or two phases and make the power system unbalanced. As a result, they are referred to as unbalanced faults.

Shunt faults can be permanent (leading to sustained outages) or temporary (leading to momentary outages). Permanent faults occur when there is permanent damage to a power system equipment and require line crew to make repairs before reenergization. Temporary faults occur due to lightning, animal contact, flying debris, and other temporary sources of fault. Such faults clear out on their own after the fault arc gets extinguished. When the arc gets extinguished by itself without the operation of any protective device, the fault is referred to as a self-clearing fault. These faults generally occur when insulation breaks down near the voltage peak but clear out on their own within a quarter cycle. The frequency of self-clearing faults increase over time and eventually lead to permanent faults [6]. Fault arcs can also get extinguished by the operation of a relay and circuit breaker. After a short open time delay, which allows the arc to get extinguished, the relay sends a reclose command to the circuit breaker to resume normal operation. Most faults on underground cables are permanent faults. In contrast, most faults on overhead systems are temporary faults [7].

Shunt faults can cause significant thermal damage to power system equipment. Thermal energy during a fault is proportional to the magnitude and duration of the fault current. If this thermal energy exceeds the thermal limit of power transformers, motors, and other power system equipment, they get damaged due to insulation failure. In addition, strong mechanical forces developed by the high magnitude current can break and physically damage power system equipment. In fact, [8] reports that transformers, a critical asset in the substation, most often fail due to mechanical and thermal stress caused by external through faults. Shunt faults are also a safety concern. Sparks from faults can start forest fires. Faults inside oil-filled transformers can lead to fires and explosions, creating dangerous working conditions for personnel inside the substation. Arc flash events in a switchgear can lead to dangerous and possibly fatal conditions due to heat, ultraviolet radiation, shrapnel, noise, and pressure from the blast [9]. Shunt faults if not cleared before the critical clearing time can make the power system unstable, leading to cascading outages [10]. Finally, shunt faults can cause voltage sags or swells on other healthy feeders. A voltage sag is defined as an event in which the rms voltage drops to a value between 0.1 and 0.9 per unit for a duration

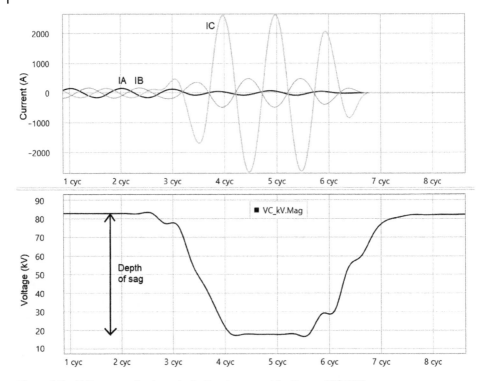

Figure 1.3 Voltage sag due to a single line-to-ground fault on a 138 kV line.

between a half cycle to one minute. An example of voltage sag during a single line-to-ground fault is shown in Fig. 1.3. Voltage sag is a power quality event that can shut down sensitive equipment in industrial plants, resulting in a revenue loss of several million dollars. Voltage swell is defined as an event in which the rms voltage increases to a value between 1.1 and 1.8 per unit for a duration between a half cycle to one minute. This can occur when single line-to-ground faults occur on an ungrounded system. Voltage of the unfaulted phases swells to 1.73 per unit and stresses the insulators. For all the reasons listed above, shunt faults must be detected and isolated as fast as possible. The latest generation of protective relays can detect faults in as fast as 2 ms [11]. The breaker takes an additional two or three cycles to open. The fast clearing time limits the damage to power system equipment, reduces the impact of the fault on the rest of the power system, and increases personnel safety.

1.2 What Causes Shunt Faults?

This section discusses the most common causes of shunt faults on the power system.

Lightning

Lightning is a major cause of shunt faults on the power system. Data collected by Texas Reliability Entity shown summarized in Fig. 1.4 is a case in point. You can see that lightning

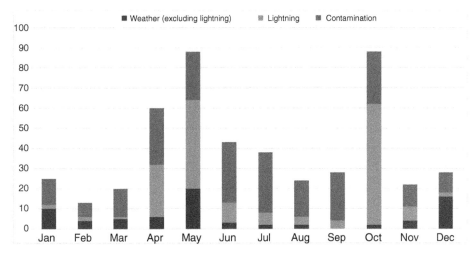

Figure 1.4 Shunt faults on 345 kV transmission lines in Texas in 2015 categorized by root cause [18].

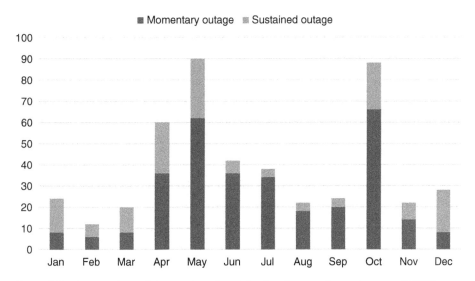

Figure 1.5 A comparison of the number of sustained and momentary outages on 345 kV transmission lines in Texas in 2015 [18].

was responsible for a large number of faults on 345 kV transmission lines in Texas in 2015, particularly during the months of April, May, and October. Figure 1.5 shows how many of the same shunt faults resulted in sustained and momentary outages. During the months of April, May, and October, when lightning activity had dramatically increased, the number of momentary outages was much greater than the number of sustained outages. The data establishes a strong correlation between lightning and momentary outages, indicating that most faults due to lightning are temporary faults.

Figure 1.6 Location of NLDN sensors to detect lightning [14].

Lightning activity in the US is monitored by the National Lightning Detection Network (NLDN). This network, owned and operated by Vaisala, has over a hundred sensors across the US as shown in Fig. 1.6. During a lightning strike, sensors that get triggered send the information they captured about the strike via satellite to a central location in Tucson, Arizona. There, data from different sensors are combined to establish date, time, peak current magnitude, type, and location of the lightning strike [13]. The data is then made available to the National Weather Service and utilities that subscribe to this service. These utilities can make use of this information to determine whether lightning was in the area when a particular fault occurred on their system and whether it was the root cause of the fault.

Lightning occurs when a thundercloud (also referred to as a cumulonimbus cloud) develops areas of positive and negative charges as shown in Fig. 1.7. Lightning discharge can occur between the positively charged and the negatively charged regions inside the cloud and are referred to as intracloud lightning. It can also occur between the positively charged

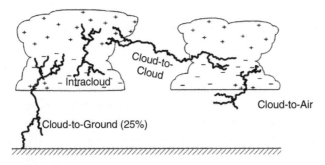

Figure 1.7 Lightning discharges can be intracloud, cloud-to-cloud (intercloud), cloud-to-air, and cloud-to-ground [14].

area of one thundercloud and the negatively charged area of another thundercloud and is referred to as cloud-to-cloud or intercloud lightning. Lightning can also occur between a thundercloud and air, referred to as a cloud-to-air lightning discharge, and between a thundercloud and the earth, referred to as a cloud-to-ground discharge. Cloud-to-ground discharges constitute about twenty-five percent of all lightning discharges and are the ones responsible for creating shunt faults on the power system.

Figure 1.8 shows four different ways by which a cloud-to-ground lightning strike can occur. A downward negative lightning strike starts when negative charges at the lower parts of the cloud start ionizing the air. This results in a column of negative charges, known as the leader, moving down toward the ground. As the leader approaches the ground, it induces streamers of opposite charges to move up from the ground. When the leader and streamer make contact with each other, the path becomes complete, a cloud-to-ground strike occurs, and a huge amount of negative charge is transferred to the ground. A downward positive lighting strike occurs the same way as the previous one, except that the leader is initiated by positive charges in the cloud. Upward lightning strikes occur when leaders are initiated by tall objects on the ground with sharp corners. When the leader reaches the cloud, a cloud-to-ground lightning strike occurs. When the leader is positive, it is called upward positive lightning. When the leader is negative, it is called upward negative lightning. Out of the four types, downward negative cloud-to-ground lightning strikes are more common.

Figure 1.8   Types of cloud-to-ground lightning [14].

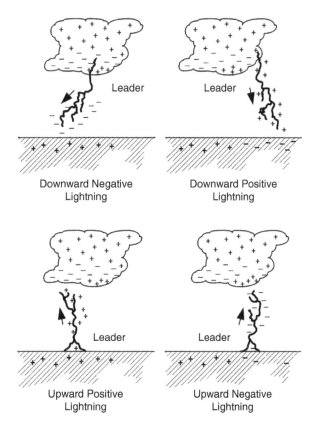

Downward Negative Lightning

Downward Positive Lightning

Upward Positive Lightning

Upward Negative Lightning

Figure 1.9 Overvoltage due to lightning strike on a 69 kV transmission line resulted in a flashover and a phase-to-ground fault. The insulator string was found to be damaged with multiple ceramic disks missing. (*Photo: Courtesy of Mr. Genardo Corpuz, Lower Colorado River Authority, USA.*)

A cloud-to-ground lightning strike can create faults when it directly strikes a phase conductor and injects a huge current surge into the line. The current surge is accompanied by a voltage surge. If the voltage surge exceeds the insulator critical flashover voltage, a flashover will occur, resulting in a shunt fault. Lightning can also strike a tower and create faults. When the injected current surge travels through the tower to the ground, a voltage rise develops across the tower crossarm due to the surge impedance of the tower and the tower footing resistance. If the voltage rise is large enough, a flashover will occur from the tower to the conductor across the insulator string. This flashover, commonly referred to as backflash, will create a shunt fault. Figure 1.9, Figure 1.10, and Figure 1.11 show the damage to utility assets due to flashover during lightning.

To protect power system equipment from lightning at substations, lighting masts (shown in Fig. 1.12) and lightning arresters are used. Line arresters are used in parallel with transmission line insulators to prevent the voltage from increasing beyond the line insulation level. To reduce the possibility of direct lightning strikes to transmission lines, ground wires (shield wires) are often placed above the phase conductors (shown in Fig. 1.13). Shield wires also reduce the possibility a backflash as the injected current is divided into three parts (tower and each direction on the shield wire). Pole ground is improved by driving a metal rod further down into the ground.

Animals

Animals such as birds, snakes, monkeys, cats, and squirrels are notorious for creating faults on the power system. Figure 1.14 shows a monkey that died after making contact with

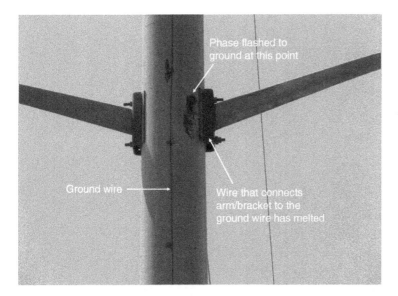

Figure 1.10 Another example of a fault due to lightning strike on a 138 kV transmission line. The resulting overvoltage caused a flashover and an A-G fault. Notice the damage to the concrete tower at the point of the flashover. Part of the wire that connects the bracket/tower arm to the ground wire has either melted or blown away. (*Photo: Courtesy of Mr. Genardo Corpuz, Lower Colorado River Authority, USA.*)

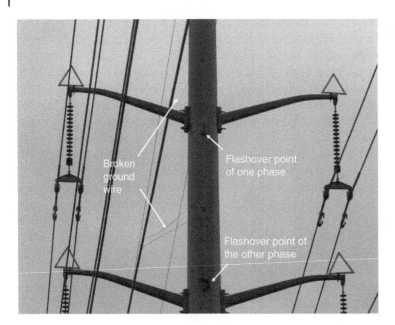

Figure 1.11 Similar to the previous example, lightning struck a 138 kV transmission line. Phase A and phase C flashed to the ground wire and created a line-to-line fault through the ground wire. Notice the evidence of damage to the concrete tower when one of the phases flashed over to the ground wire. (*Photo: Courtesy of Mr. Genardo Corpuz, Lower Colorado River Authority, USA.*)

Figure 1.12 Lightning mast at a substation.

Figure 1.13 Shield wire on a single-circuit 345 kV transmission line. (*Photo: Courtesy of Mr. Genardo Corpuz, Lower Colorado River Authority, USA.*)

Shield wire →

energized equipment in a substation and creating a fault. Figure 1.15 shows the damage to a 22 kV breaker when a bird flapped its wings, touched both B and C phases, and created a BC fault. Birds build nests on transmission towers, on distribution poles, and in substations. These nests can cause a short circuit by making contact with multiple conductors. They can also attract other animals such as snakes and raccoons that in turn cause faults. Bird droppings contaminate insulators and can result in a flashover. Woodpeckers can cause structural damage to wooden poles. Horses, bears, bison, and cattle can also degrade the structural integrity of poles by rubbing against guy-wires, causing the poles to lean and conductors to sag. Squirrels can climb utility poles and create faults by bridging the gap between phase conductors and the grounded equipment. Small animals such as rats and mice can chew on the insulation of underground cables and create faults. To reduce the number of

Figure 1.14 A monkey that made contact with energized equipment in a substation and created a fault. (*Photo: Courtesy of Mr. Emmanuel Raubenheimer, Eskom, South Africa.*)

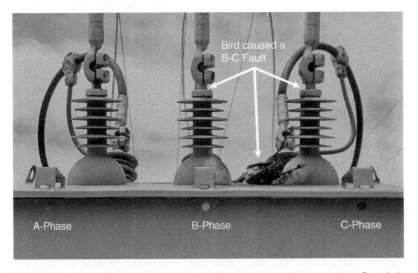

Figure 1.15 A bird flapped its wings and bridged the gap between phase B and phase C bushing on the bus side of a 22 kV breaker, creating a BC fault. The fault current magnitude was about 2 kA. The line-to-line fault later evolved into a three-phase fault. It also flashed across the breaker and created a B-G fault on the 22 kV distribution feeder, just in front of the breaker. Flash marks are clearly visible on the clamps and on the bushings of the breaker. (*Photo: Courtesy of Mr. Emmanuel Raubenheimer, Eskom, South Africa.*)

Figure 1.16 Bird spikes to prevent birds from perching.

faults due to birds, some utilities install bird spikes such as the one shown in Fig. 1.16 to discourage birds from perching or roosting. Another preventive step is to wash the insulators at periodic intervals to remove bird droppings and other contaminants. Animal guards such as the one shown in Fig. 1.17 are also installed around energized equipment to restrict animal contact.

Figure 1.17 Animal guard around the top of the transformer low voltage bushing.

Trees

Trees are responsible for a large number of power system faults. In fact, a tree was responsible for the fault that triggered the Northeast blackout of 2003, one of the major blackouts in North America's history [15]. Trees can cause faults in a number of ways. Those uprooted by heavy winds or hurricanes can tear down lines, knock down poles, or damage insulators when falling down. Overgrown vegetation can bridge phase conductors. Tree limbs broken during heavy winds can fly into a line and bridge phase conductors. Tree branches blown by the wind can push two conductors together. Lines may sag during heavy load conditions and on doing so may make contact with the underlying vegetation. Electric utilities typically have a vegetation management program to trim trees and prevent tree-related outages and wildfires. They may also modify spacing between phase conductors to increase the resistance to flashover.

Other Causes

Power system faults can be caused by accidents such as when vehicles crash into poles or when drones, kites, shiny foil balloons, and hot air balloons make contact with energized conductors. Such an unfortunate event occurred on July 30, 2017, when a hot-air balloon struck high-voltage conductors, killing all sixteen people aboard [16]. A fault can also occur due to human errors such as forgetting to remove the grounding chains after maintenance and closing the breaker. Faults can also occur during acts of vandalism. They include thieves stealing conductor wire to later sell as scrap metal or people shooting at insulators with a rifle. An example of vandalism occurred on April 16, 2013, when there was a sniper attack on the Metcalf transmission substation owned by Pacific Gas and Electric [17]. Snipers opened fire on seventeen transformers, causing severe damage and forcing grid officials to reroute power from nearby power plants to avoid a blackout. In addition, contaminants can also weaken the insulators over time, resulting in a flashover and a short-circuit fault. This is a problem particularly in coastal areas where contaminants such as dust and salt or in agricultural areas where contaminants such as pesticide and fertilizer build up over time, eventually leading to a fault. Strong winds can cause power lines to swing into one another and create faults. Severe winds during hurricanes and tornadoes can even break power lines and utility poles, creating significant damage to the power system infrastructure. Snow and ice are also a major cause of faults. Their weight can cause power lines to snap or tree limbs to break and fall into utility lines. Power system equipment can also fail internally and create a fault. The internal fault may be the result of an insulation failure due to age, overvoltage, and other factors. Figure 1.18 show when a regulator failed in a substation and caused significant damage. Figure 1.19 shows another example when there was a fault inside a capacitor bank due to overvoltage.

Figure 1.18 Internal fault within a voltage regulator at a substation. (*Photo: Courtesy of Mr. Long Tran, South Texas Electric Cooperative, USA.*)

Figure 1.19 Internal fault inside a capacitor bank in a substation due to overvoltage. (*Photo: Courtesy of Mr. Long Tran, South Texas Electric Cooperative, USA.*)

1.3 Aim and Importance of Fault Location

The purpose of the power system is to deliver electric power from generators in power plants to industrial plants and residential customers through an interconnected network of transmission lines and distribution feeders. Figure 1.20 shows a turbine and generator inside a steam power plant. Steam produced by heating water is used to rotate the turbine. This turbine coupled to an electric generator transforms the mechanical power to electrical power with a voltage between 5 and 34.5 kV. To transport electrical power over long distances with minimum loss, the voltage is stepped up to between 69 and 765 kV using step-up transformers as shown in Fig. 1.21. The electrical power at a stepped-up voltage is

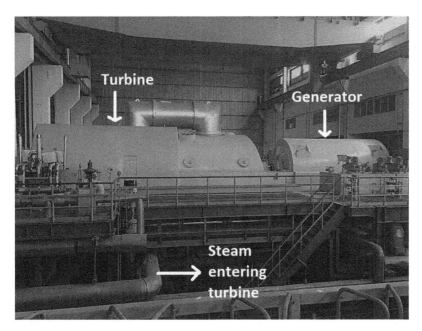

Figure 1.20 Turbine and generator in a 125 MW steam power plant. (*Photo: Courtesy of Mr. Dip Kumar Das, Development Consultants Private Limited, India.*)

Figure 1.21 Generator step-up transformer is transforming the low voltage output from the steam turbine generator to 132 kV and interconnecting the power station to the transmission network. (*Photo: Courtesy of Mr. Dip Kumar Das, Development Consultants Private Limited, India.*)

then transported by transmission lines. Closer to load centers, the voltage is stepped down to the low-voltage level of the load and delivered to various individual users.

When a permanent fault occurs, this normal flow of power is interrupted. Loads downstream from the fault experience an outage. Unless power is rerouted through other lines, the outage time will equal the time taken to find the fault plus the time taken to make repairs. If power is rerouted, the other lines supporting the additional load may get overloaded since power systems today operate close to their operating limits. This can set off a series of cascading trips, leading to a blackout. Data collected by NERC shows that between 2015 and 2019, 200 kV circuits had an average unavailability rate of 0.09% due to automatic outages caused by protective relay operation during an unplanned transmission incident (see Figure 1.22). This means that at any given time, there was a 0.09% chance that a 200 kV transmission circuit was unavailable due to a sustained automatic outage [12]. Figure 1.23 shows the outage data by duration in Texas [18]. As you can see, fifty-eight percent of the outages lasted longer than two hours. As a result, it is critical to find permanent faults as quickly as possible and restore normal operation. In addition to permanent faults, it is equally important to locate temporary faults. While they may not have resulted in a sustained outage, locating them and taking immediate action can help the utility avoid a permanent fault and outage later.

Faults inside generating power plants, substations, or load centers are localized and are hence easier and quicker to locate. In contrast, faults on transmission and distribution lines are much more difficult to find as they cover large distances. Reference [19] reports that there are 707,000 miles of high-voltage transmission lines and 6.5 million miles of distribution lines in the US. In the aftermath of a fault, the traditional approach for fault location has been to patrol the line length and look for visible evidence of a fault.

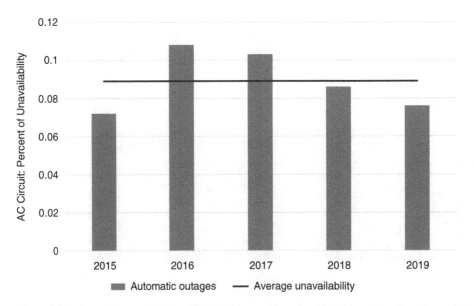

Figure 1.22 Percent unavailability of 200 kV transmission circuits due to automatic outages from 2015 to 2019 [12].

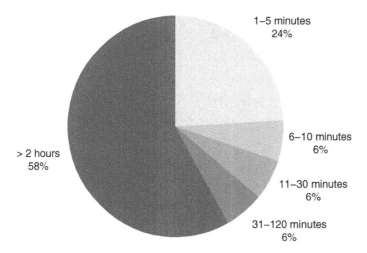

Figure 1.23 2015 345 kV automatic outage data by duration in Texas [18].

This manual approach is labor intensive and time consuming. Calls from customers reporting an outage can sometimes help narrow down the location of the fault. To make matters worse, rough terrains, remote areas, or bad weather may interfere with line patrol efforts and can delay repair work. Utilities may be forced to consider other expensive alternatives such as deploying helicopters to locate the fault. Furthermore, faults that leave behind a trail of evidence are easy to find. However, some faults are difficult to find. Figure 1.24 shows one such scenario where a flashed insulator is not easily visible from the ground. Figure 1.25 shows a zoomed-in view of the blown-out insulator. In addition, temporary faults may not leave a visible mark or indication as to the cause and location of the fault and can go unnoticed during inspection patrols. Knowing where to look for the fault can save field crew significant time and help them restore power back quickly. In addition to helping field crew locate the fault, fault location provide valuable operational benefits. For example, knowing the fault location is critical to establishing whether a protective relay operated correctly or not during a fault. It can also help catch CT and PT wiring errors and help validate relay settings, all of which are important to avoid undesired relay operations. These additional benefits are discussed in detail in Chapter 7.

1.4 Types of Fault-Locating Algorithms

The subject of using electrical signals for fault location has been of interest since the early 1900s [20–22]. Today, fault-locating algorithms can be classified into two major categories, impedance-based and traveling-wave algorithms.

Impedance-based fault-locating algorithms use Ohm's law to calculate fault location. Voltage during the fault is divided by current during the fault to estimate the line impedance between the measurement point and the shunt fault. Line impedance to the fault is not as useful as knowing the distance to the fault in miles (or kilometers). Therefore, these

Figure 1.24 Faults are not always immediately obvious during line patrol. For example, an insulator flashover at this structure resulted in a phase-to-ground fault. The flashed insulator is not visible from the ground. (*Photo: Courtesy of Mr. Genardo Corpuz, Lower Colorado River Authority, USA.*)

algorithms use the line impedance in ohms per unit length (length in miles or kilometers) to convert line impedance to the fault to a distance estimate (same unit as the line length). Since line impedance is usually defined at the system frequency, the voltage and current used for fault location are also at the system frequency. Impedance-based fault-locating algorithms can be applied to locate faults on both transmission and distribution lines.

Traveling-wave fault-locating algorithms, on the other hand, use the velocity equation to calculate fault location. A fault causes a sudden change in voltage at the fault point. This generates high frequency voltage and current surges, also known as traveling waves, which travel at either direction from the fault. These waves travel at almost the speed of light on

Figure 1.25 A zoom-in view of the flashed and blown-out insulator. (*Photo: Courtesy of Mr. Genardo Corpuz, Lower Colorado River Authority, USA.*)

overhead lines and eventually die out due to multiple reflections. Multiplying the velocity of the waves with the time taken by the waves to travel from the fault point to the measurement point gives the distance to the fault from the measurement point. Traveling-wave fault-locating algorithms can be presently used to locate faults on transmission lines only (69 kV and above).

In addition to impedance-based and traveling-wave algorithms, artificial neural networks, fuzzy logic, and other knowledge-based approaches are also being studied extensively for fault location [23–25]. Since this category is in the research stage and not commercially available, this book will focus on impedance-based and traveling-wave algorithms.

1.5 How are Fault-Locating Algorithms Implemented?

Fault-locating algorithms are implemented in microprocessor-based relays, digital fault recorders, stand-alone fault locators, and fault analysis software. In this section, we discuss their implementation.

(a) Microprocessor-Based Protective Relays

There have been significant advancements in the field of protective relaying. The earliest relays were electromechanical relays. They used the measured voltage, current, or both to create electromechanical forces. During a fault, these forces would actuate a movable

Figure 1.26 Electromechanical relay for overcurrent protection.

element within the relay to close the trip output contact. A target on the relay would indicate when the relay tripped. Other than this, the relay did not provide any other information about the fault. Figure 1.26 shows an example electromechanical relay designed for overcurrent protection.

Today, most protective relays are microprocessor-based relays, also referred to as numerical or digital relays. These relays use analog-to-digital converters to digitize the current and voltage analog signals. The digitized data is then used by numerical protection algorithms saved inside the microprocessor to detect faults and take action by closing the relay trip output contact. Microprocessor-based relays are multifunctional devices, meaning the relay is capable of providing several protection functions. For example, the microprocessor-based relay shown in Fig. 1.27 is a feeder protection relay. It can provide overcurrent, under- and overvoltage, under- and overfrequency, directional, breaker failure, reclosing, and other protective functions. In addition to protecting the power system, microprocessor-based relays also capture event records during a fault and provide fault location to assist operators understand the fault, analyze the relay operation, and restore service quickly. References [26–28] discuss all the advantages of using microprocessor-based relays over electromechanical relays. For the purposes of this book, we will focus on the fault location feature of microprocessor-based relays.

Fault location is a standard feature available in most microprocessor-based relays. An impedance-based fault-locating algorithm is used to calculate fault location from event reports saved after a fault. Figure 1.28 shows an example event report captured during a

Figure 1.27 Microprocessor-based multifunctional feeder protection relay and fault locator. (*Photo: Courtesy of Schweitzer Engineering Laboratories.*)

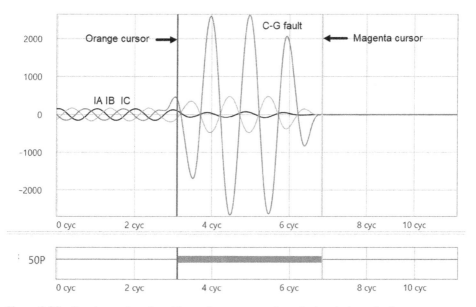

Figure 1.28 Event report captured by a microprocessor-based relay during a fault.

fault. It is a snapshot of the measured voltage and current analog signals, as well as the response of relay protection elements for a length of time set by the user (eleven cycles in this example). The fault-locating algorithm in the relay first establishes the window of time for which the fault was present on the power system. One method to establish this window is to look at the response of overcurrent, distance, and other fault detector elements in the relay during the fault [29]. For instance, in the example event report, the phase instantaneous overcurrent element (50P) picked up at the location of the orange cursor. It dropped out at the location of the magenta cursor. The relay interprets the orange cursor to be the start of the fault and the magenta cursor to be the end of the fault. Data between these two cursors is, therefore, valid for fault location. Since currents are most stable in the middle of the window, fault-locating algorithms in these relays use voltage and current data in the middle of the window to calculate fault location.

In addition to impedance-based fault location, some microprocessor-based relays also implement traveling-wave fault location. They have special hardware to capture traveling waves measured by conventional CTs and PTs. Traveling waves are not just created by internal faults on the line. They are created by anything that causes a sudden change in voltage such as lightning strikes or switching operations. Therefore, relays implementing traveling-wave fault location first determine if the captured traveling wave is due to an internal fault on the line. They usually do this by looking for a trip output from the relay within a couple of cycles from the instant the first traveling wave was recorded. This is because internal faults on the line will cause the relay to trip [33].

(b) Digital Fault Recorders (DFRs)

Digital fault recorders (DFRs) are microprocessor-based devices that record three-phase voltage and current waveforms, as well as the sequence of events following a disturbance on the power system. The recording capabilities of a DFR may seem similar to a microprocessor-based protective relay. However, there are some differences between the two with respect to sampling rate and how long of a record they save [30]. DFRs typically record current and voltage waveforms at a higher sampling rate than most protective relays. For example, [31] records waveforms at 256 samples per cycle and [32] records waveforms at 200 samples per cycle. In contrast, a typical relay records data at 4, 16, or 32 samples per cycle although the latest generation of microprocessor-based relays record data at a resolution of 1 MHz [33]. The higher sampling rate allows the visualization of high frequency transients such as lightning strikes, breaker restrikes, and switching events. DFRs also record data over a longer length of time (typically several minutes) than most protective relays (typically several cycles). In addition to recording waveforms during a fault, some DFRs can also locate faults. For example, [32] locates faults using impedance-based algorithms.

(c) Stand-Alone Fault Locators

Stand-alone fault locators provide a fault location after a line trip. An example of such a device has been described in [34]. It is a traveling-wave fault locator that can provide fault location for eight different circuits. Installing this particular fault locator does not require a line outage. It uses split-core linear couplers that are placed around the secondary wiring of conventional CTs to capture the traveling waves for fault location. The trip output from a protective relay can be wired into the digital input of the fault locator. A change of state of the digital input would indicate an internal fault on the line and would trigger the fault locator.

(d) Fault Analysis Software

Fault analysis software such as [35] automatically collects waveform event reports from microprocessor-based protective relays or DFRs after a fault, runs fault location algorithms, and notifies personnel about an event along with fault location results via email or SMS text messages.

1.6 Evaluation of Fault-Locating Algorithms

Fault-locating algorithms are evaluated by their error percent. This is typically calculated relative to line length as [36]:

$$e = \frac{|m_{est} - m_{act}|}{LL} \times 100 \tag{1.1}$$

where m_{est} is the fault location estimate, m_{act} is the actual fault location, and LL is the line length. All three must have the same unit of distance, typically miles or kilometers. A lower error percent is desirable as this means that the estimated fault location is closer to the actual fault location. This allows field crew to find the fault faster, leading to shorter outage times. In addition to measuring the performance of fault-locating algorithms, knowing the error percent of a specific fault-locating algorithm for a particular line can help field crew build a search radius around the fault location estimate. For example, suppose that a fault has occurred on a hundred-mile-long line. The fault is estimated to be located at forty-five miles from the substation. Past experience indicates that the fault-locating error on this line is five percent of the line length. The error works out to be five miles in this example. Therefore, field crew should broaden their search around the estimated fault location by five miles.

Commercial fault-locating devices may also be evaluated on how fast they can calculate fault location. Traditionally, the purpose of fault-locating algorithms was to help field crew search for the fault. Therefore, it was perfectly acceptable if this data was available within a couple of minutes after the fault had occurred. Some applications, however, use the fault location estimate to make protection decisions. An example of such an application is the decision to autoreclose on hybrid lines. Hybrid line consists of both overhead and underground line sections. Faults on overhead line sections can be momentary or permanent while faults on underground line sections are typically permanent. Protective relays can use the fault location data to identify whether a fault is on the underground or on the overhead line section. If the fault is determined to be on the overhead line section, reclosing is allowed. On the other hand, if the fault is on the underground section, reclosing would be blocked. Such applications require the fault location to be computed and be available faster than that required for traditional applications.

1.7   The Best Fault-Locating Algorithm

Under impedance-based and traveling-wave categories, several fault-locating algorithms have been developed. While all algorithms have the same objective, which is to find the fault with the highest accuracy, each of them make different assumptions and use different data to arrive at this result. What is the best fault-locating algorithm? Unfortunately, there is no one-size-fits-all answer. The correct answer is that it depends. It depends on the data available for fault location, the system to which the algorithm will be applied to, and the characteristics of the fault (such as whether it involves the ground or whether it evolves from one fault type to the other). The purpose of this book is to explain the theory and principles behind many of the common fault-locating algorithms, identify the input data they need, and discuss error sources. This will allow the user to make an informed decision about the best fault-locating algorithm for their system.

1.8   Summary

The key takeaways from this chapter are as follows:

- The power system can experience two types of faults, series faults and shunt faults.
- Series faults occur when there is an open circuit on one or two phase conductors during load conditions. The current in the faulted phase decreases due to loss in load while the current in the healthy phase (or phases) equal the load current. This unbalance can damage transformers and motors. Series faults are caused by blown fuses or broken jumpers on one or two phases. They can also be caused when one or two poles of a circuit breaker are unable to close during a manual or automatic close operation.
- Shunt faults occur when there is a shunt connection between one or more phase conductors to the ground or between each other. The current in the faulted phase dramatically increases while the voltage of the faulted phase sags. These faults need to be detected and isolated as fast as possible to limit the thermal and mechanical stress to power system equipment, prevent the power system from becoming unstable, avoid shutting down industrial plants with equipment sensitive to voltage sags, and increase personnel safety. Shunt faults are caused by lightning, animals, tree contact, equipment failure, extreme winds or snow, vehicle accidents, vandalism, or human error. They can be single line-to-ground, line-to-line, double line-to-ground, or three-phase faults. Single line-to-ground faults are the most common fault type.
- Fault-locating algorithms are used to locate shunt faults on transmission and distribution lines to expedite service restoration and repair after a permanent fault. They can be classified into impedance-based and traveling-wave algorithms. These algorithms can be implemented in microprocessor-based relays, digital fault recorders, stand-alone fault locating devices, or fault analysis software.
- Fault-locating algorithms are required to be accurate. In some applications where the fault location information is used to make a protective decision such as reclosing, how fast fault location is calculated becomes another important requirement.
- Under impedance-based and traveling-wave categories, several algorithms have been developed. The best method for fault location depends on your application, the available data, and the characteristics of the fault.

2

Symmetrical Components

In Chapter 1, we learned about the different types of shunt faults on the power system. When setting protective relays or sizing circuit breakers, it is important to know the short-circuit current magnitudes during a fault. Currents during a three-phase fault are easy to calculate. Only a single phase needs to be solved to calculate the current of that phase during the fault. Currents in the other two phases have the same magnitude and a phase angle shift of 120 degrees since three-phase faults are balanced faults. While the single-phase approach simplifies the analysis of balanced faults, it cannot be applied to the analysis of unbalanced faults such as single line-to-ground, line-to-line, or double line-to-ground faults. This is when the tool of symmetrical components is used.

Symmetrical components is a mathematical tool developed in 1913 by Charles L. Fortescue when investigating the behavior of an induction motor in an n-phase power system during unbalanced conditions [37]. Contributions from C. F. Wagner, R. D. Evans, W. A. Lewis, E. L. Harder, and E. Clarke extended the application of symmetrical components to the analysis of unbalanced faults on a three-phase power system. In addition to analyzing unbalanced faults, symmetrical components have other applications. They are used to commission digital protective relays and identify wiring errors in the CT and PT connections to the relay, as well as errors in the phase rotation setting of the relay. They are also used as an input to relay protection algorithms in digital relays to make tripping decisions during a fault and to fault-locating algorithms. As a result, it is important to have a thorough understanding of symmetrical components.

The subject of symmetrical components is quite extensive. This chapter focuses on discussing the important concepts that are required to understand the derivation of fault-locating algorithms in later chapters. For a more in-depth review of symmetrical components, refer to [38, 39, 41–44]. Since symmetrical components is applied to phasors, we begin the chapter by reviewing phasors. Next, we introduce the concept of symmetrical components and show how to transform phasors to symmetrical components and vice versa. We then discuss sequence networks and how to interconnect them during the different shunt faults. Finally, we explain how to calculate sequence impedances of a three-phase line. The sequence line impedances are used as an input to impedance-based fault-locating algorithms.

Fault Location on Transmission and Distribution Lines: Principles and Applications, First Edition.
Swagata Das, Surya Santoso, and Sundaravaradan N. Ananthan.
© 2022 John Wiley & Sons Ltd. Published 2022 by John Wiley & Sons Ltd.
Companion website: www.wiley.com/go/das/faultlocation

2.1 Phasors

A phasor is a complex number that is used to represent a sinusoidal signal in terms of its magnitude and phase angle. A sinusoidal signal is one whose frequency, amplitude (or magnitude), and phase angle does not change with time. Voltage and current in AC electrical circuits are sinusoidal signals during steady-state conditions (see Fig. 1.3). Mathematically, a sinusoidal signal can be represented by a sine or a cosine function. In this example, we will use the cosine function as shown below.

$$v(t) = V_{max} \cos(\omega t + \theta), \tag{2.1}$$

where $v(t)$ is a time-domain sinusoidal voltage signal, ω is the angular frequency of the voltage signal in radians per second, V_{max} is the peak magnitude of the voltage signal in volts, and θ is the phase angle of the voltage signal with respect to a reference in radians. When analyzing AC electrical circuits, we typically represent the magnitude of voltage or current in rms instead of peak. Therefore, (2.1) can be written in terms of its rms value as shown below.

$$v(t) = \sqrt{2} \times V_{rms} \cos(\omega t + \theta) \qquad \because V_{rms} = \frac{V_{max}}{\sqrt{2}}. \tag{2.2}$$

To transform the time domain signal to the complex domain, Euler's formula shown below is used.

$$e^{jx} = \cos x + j \sin x. \tag{2.3}$$

Substituting (2.3) in (2.2), we get:

$$v(t) = \Re \left\{ \sqrt{2} \times V_{rms} \, e^{j(\omega t + \theta)} \right\}$$

$$= \Re \left\{ \underbrace{V_{rms} \, e^{j\theta}}_{\text{phasor}} \times \sqrt{2} \, e^{j\omega t} \right\} \qquad \because e^{j(\omega t + \theta)} = e^{j\omega t} \times e^{j\theta}, \tag{2.4}$$

where $\Re$ stands for the real part of a complex number. Equation (2.4) shows that a sinusoidal signal in the time domain can be written as the product of two complex numbers in the complex domain. The complex number that is independent of time is known as a phasor. It has a magnitude equal to the rms value of the signal and an angle equal to the phase angle of the signal. The other complex number $\sqrt{2} \, e^{j\omega t}$ has a magnitude of $\sqrt{2}$ and is a function of time. As time increases, it rotates at an angular velocity of ω in the counterclockwise direction (by convention) and completes one revolution in one cycle. The phasor shown in (2.4) has been written in the exponential form. Phasors can also be written in the rectangular and exponential form as:

$$V = V_{rms} \cos \theta + jV_{rms} \sin \theta \quad \text{(rectangular)}$$

$$= V_{rms} \angle \theta \qquad \qquad \text{(polar)} \tag{2.5}$$

The rectangular form is useful when adding or subtracting two phasors. The polar form is useful when multiplying or dividing two complex numbers. Phasors are widely used as they simplify the analysis of AC electrical circuits during steady state.

2.2 Theory of Symmetrical Components

The method of symmetrical components takes an unbalanced set of three-phase voltage and current phasors and decomposes them into three sets of balanced phasors known as sequence or symmetrical components. The three sets are zero-sequence, positive-sequence, and negative-sequence components. This is illustrated in Figure 2.1. The unbalanced three-phase current phasors have an ABC phase sequence. (Phase sequence refers to the order in which the phasors cross a stationary point when rotating in the counterclockwise direction.) Zero-sequence components have the same magnitude with zero phase shift between the three phases. Positive-sequence components have the same magnitude, the same phase sequence as the system phase sequence (ABC in the illustration), and a phase shift of 120 degrees between the three phases. Negative-sequence components have the same magnitude, a phase sequence that is opposite to the system phase sequence (ACB in the illustration), and a phase shift of 120 degrees between the three phases.

Mathematically, the three-phase currents can be written in terms of their symmetrical components as:

$$I_a = I_{a0} + I_{a1} + I_{a2}$$
$$I_b = I_{b0} + I_{b1} + I_{b2} \tag{2.6}$$
$$I_c = I_{c0} + I_{c1} + I_{c2}$$

This can be further simplified by expressing the sequence components for the B- and C-phases in terms of the A-phase. Because zero-sequence components are equal to each other in magnitude and phase angle, we can replace I_{b0} and I_{c0} in the above equations with I_{a0}. To write I_{b1} and I_{c1} in terms of I_{a1}, and I_{b2} and I_{c2} in terms of I_{a2}, we use an operator a that has a magnitude of one and a phase angle of 120 degrees. Since this operator has a unit magnitude, multiplying a phasor with a will simply rotate the phasor by 120 degrees

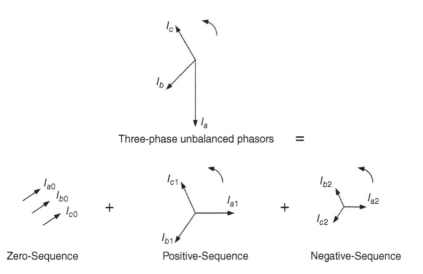

Three-phase unbalanced phasors $=$

Zero-Sequence $\quad+\quad$ Positive-Sequence $\quad+\quad$ Negative-Sequence

Figure 2.1 Unbalanced three-phase current phasors decomposed into three sets of balanced phasors. Phasors always rotate in the counterclockwise direction.

in the counterclockwise direction (phasors always rotate in the counterclockwise direction as discussed in Section 2.1). Similarly, multiplying a phasor with a^2 will simply rotate the phasor by 240 degrees in the counterclockwise direction. Another expression that is useful to remember and will be used later is:

$$1 + a + a^2 = 0 \tag{2.7}$$

Using a, we can write the below relations:

$$
\begin{aligned}
I_{b1} &= a^2 I_{a1}, \\
I_{c1} &= a I_{a1}, \\
I_{b2} &= a I_{a2}, \\
I_{c2} &= a^2 I_{a2}.
\end{aligned}
\tag{2.8}
$$

Substituting (2.8) in (2.6), we get:

$$
\begin{aligned}
I_a &= I_{a0} + I_{a1} + I_{a2}, \\
I_b &= I_{a0} + a^2 I_{a1} + a I_{a2}, \\
I_c &= I_{a0} + a I_{a1} + a^2 I_{a2}.
\end{aligned}
\tag{2.9}
$$

To further simplify, let us drop the a subscript from the symmetrical components and write the above equation in a more compact fashion as:

$$
\begin{bmatrix} I_a \\ I_b \\ I_c \end{bmatrix} = \underbrace{\begin{bmatrix} 1 & 1 & 1 \\ 1 & a^2 & a \\ 1 & a & a^2 \end{bmatrix}}_{\text{Transformation matrix}} \times \begin{bmatrix} I_0 \\ I_1 \\ I_2 \end{bmatrix}
\tag{2.10}
$$

$$
\begin{aligned}
\therefore \begin{bmatrix} I_0 \\ I_1 \\ I_2 \end{bmatrix} &= \begin{bmatrix} 1 & 1 & 1 \\ 1 & a^2 & a \\ 1 & a & a^2 \end{bmatrix}^{-1} \times \begin{bmatrix} I_a \\ I_b \\ I_c \end{bmatrix} \\
&= \frac{1}{3} \begin{bmatrix} 1 & 1 & 1 \\ 1 & a & a^2 \\ 1 & a^2 & a \end{bmatrix} \times \begin{bmatrix} I_a \\ I_b \\ I_c \end{bmatrix}
\end{aligned}
\tag{2.11}
$$

Equation 2.10 shows how to calculate three-phase phasors from symmetrical components. Equation 2.11 shows how to calculate symmetrical components from three-phase phasors. Even though the a subscript was dropped, it is understood that symmetrical components in the above equations are for the A-phase. While (2.10) and (2.11) show the transformation for the currents, you can do the same with voltages as well. When deriving (2.10) and (2.11), two assumptions were made. First, the system was assumed to have an ABC phase sequence. If the system had ACB phase rotation instead, it is necessary to swap I_b and I_c in the two equations. Second, the symmetrical components for the B- and C-phases were written in terms of the A-phase. Therefore, A-phase is the reference phase in (2.10) and (2.11). During a balanced condition, any of the three phases can be chosen to be the reference phase. Typically A-phase is chosen as the reference. However, during an unbalanced fault, the reference phase should be selected based on the fault type as explained in

Section 2.3. The reference phase should be placed on top in (2.10) and (2.11). The phases following the reference phase should be listed according to the system phase rotation. It is also important to point out that when the power system is perfectly balanced, only positive-sequence quantities are expected to be present. The magnitude and phase angle of the positive-sequence current and voltage will equal those of the selected reference phase.

2.3 Interconnecting Sequence Networks

A three-phase network can be represented in terms of three sequence networks, one for each of the three sequence currents to flow. Positive-sequence current flows in the positive-sequence network, negative-sequence current flows in the negative-sequence network, and zero-sequence current flows in the zero-sequence network. Refer to [39, 44] on how to create sequence networks with power system components such as generators, motors, lines, and transformers. Out of the three sequence networks, only the positive-sequence network has sources. This is because generators are designed to produce only positive-sequence voltage. When the power system is balanced, the three sequence networks are not connected with one another. As a result, only positive-sequence currents flow during balanced conditions. There is no negative-sequence or zero-sequence currents as those networks do not have a source. When there is an unbalanced condition on the power system, the negative-sequence network and, in some cases, the zero-sequence network get connected to the positive-sequence network at the point of the unbalance. The source in the positive-sequence network then drives current through all the networks. This section shows how to interconnect sequence networks when the unbalance is created by a shunt fault. In the discussion that follows, $I_{F_a}^f$, $I_{F_b}^f$, and $I_{F_c}^f$ are the phase currents at the fault point, and $V_{F_a}^f$, $V_{F_b}^f$, and $V_{F_c}^f$ are the phase voltages at the fault point.

(a) Three-Phase Faults

A three-phase fault is a balanced fault. As a result, only the positive-sequence network exists. The fault point in the positive-sequence network (F_1) is connected to the reference bus (N_1) through a fault resistance R_f as shown in Figure 2.2(a).

(b) Single Line-to-Ground Fault

Consider an A-G fault with a fault resistance of R_f between the phase and ground. This is an unbalanced fault and our task is to determine how to connect the sequence networks. For simplicity, assume that the power system was unloaded at the time of the fault (unfaulted phase currents are zero). The following relations are true at the fault point:

$$I_{F_b}^f = I_{F_c}^f = 0, \tag{2.12}$$

$$V_{F_a}^f = R_f \times I_{F_a}^f. \tag{2.13}$$

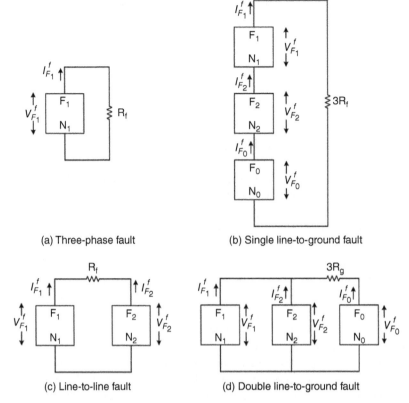

(a) Three-phase fault

(b) Single line-to-ground fault

(c) Line-to-line fault

(d) Double line-to-ground fault

Figure 2.2 Interconnection of sequence networks for the different shunt faults.

Assuming an ABC phase rotation and selecting A-phase as the reference (the importance of selecting A-phase as the reference will be discussed later), the sequence currents at the fault point are:

$$
\begin{bmatrix} I_{F_0}^f \\ I_{F_1}^f \\ I_{F_2}^f \end{bmatrix} = \frac{1}{3} \begin{bmatrix} 1 & 1 & 1 \\ 1 & a & a^2 \\ 1 & a^2 & a \end{bmatrix} \times \begin{bmatrix} I_{F_a}^f \\ I_{F_b}^f \\ I_{F_c}^f \end{bmatrix}
$$

$$
= \frac{1}{3} \begin{bmatrix} 1 & 1 & 1 \\ 1 & a & a^2 \\ 1 & a^2 & a \end{bmatrix} \times \begin{bmatrix} I_{F_a}^f \\ 0 \\ 0 \end{bmatrix}, \qquad \because I_{F_b}^f = I_{F_c}^f = 0
$$

(2.14)

$$
\therefore I_{F_0}^f = I_{F_1}^f = I_{F_2}^f = \frac{1}{3} I_{F_a}^f.
$$

Substituting (2.14) in (2.13) and writing $V_{F_a}^f$ in terms of symmetrical components, we get:

$$
V_{F_0}^f + V_{F_1}^f + V_{F_2}^f = 3 \times R_f \times I_{F_1}^f.
$$

(2.15)

From (2.14) and (2.15), we can draw three conclusions. First, all three sequence currents are present during a single line-to-ground fault. Second, the three sequence networks must be connected in series to satisfy the condition of three sequence currents being equal to each other during the fault. Third, since the sum of the sequence voltages at the fault point equals the voltage drop across the fault resistance, the fault resistance must be placed in series with the sequence networks. The interconnection of the sequence networks is shown in Figure 2.2(b). The fault point in the zero-sequence network (F_0) is connected to the reference bus (N_2) of the negative-sequence network. The fault point in the negative-sequence network (F_2) is connected to the reference bus (N_1) of the positive-sequence network. The fault point in the positive-sequence network (F_1) is connected back to the reference bus (N_0) of the zero-sequence network through a fault resistance $3R_f$ to complete the circuit. The fault resistance is shown as $3R_f$ to get the same voltage drop shown in (2.15).

The results given by (2.14) and (2.15) were used to determine how to interconnect the sequence networks during a single line-to-ground fault. Those results were obtained by assuming an ABC phase rotation and using phase A as the reference for an A-G fault. The results would have been the same for a system with ACB phase rotation. However, if any other phase was used as the reference for the A-G fault, the results would have been different and the connection of the sequence networks shown in Figure 2.2(b) would no longer be valid. Therefore, choosing the correct reference phase is important. For single line-to-ground faults, always use the faulted phase as the reference phase.

(c) Line-to-Line Fault

Figure 2.3 shows a BC fault with a fault resistance of R_f between the phases. For simplicity, assume the system was unloaded at the time of the fault (unfaulted phase current is zero). During this fault, the current in one faulted phase is equal and opposite to the current in the other faulted phase. Furthermore, at the fault point, the faulted phase voltages would be equal to each other when there is no fault resistance. In the presence of fault resistance, the voltage of one faulted phase is equal to the voltage of the other faulted phase plus an additional voltage drop due to flow of fault current through the fault resistance. These relations at the fault point are summarized below.

$$I^f_{F_a} = 0, \tag{2.16}$$

$$I^f_{F_b} = -I^f_{F_c}, \tag{2.17}$$

$$V^f_{F_b} - V^f_{F_c} = R_f I^f_{F_b}. \tag{2.18}$$

Figure 2.3   BC fault on an unloaded power system.

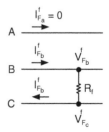

Assuming an ABC phase rotation, the sequence currents at the fault point are:

$$
\begin{bmatrix} I^f_{F_0} \\ I^f_{F_1} \\ I^f_{F_2} \end{bmatrix} = \frac{1}{3} \begin{bmatrix} 1 & 1 & 1 \\ 1 & a & a^2 \\ 1 & a^2 & a \end{bmatrix} \times \begin{bmatrix} I^f_{F_a} \\ I^f_{F_b} \\ I^f_{F_c} \end{bmatrix}
$$

$$
= \frac{1}{3} \begin{bmatrix} 1 & 1 & 1 \\ 1 & a & a^2 \\ 1 & a^2 & a \end{bmatrix} \times \begin{bmatrix} 0 \\ I^f_{F_b} \\ -I^f_{F_b} \end{bmatrix} \tag{2.19}
$$

$$
= \frac{1}{3} \begin{bmatrix} 0 \\ (a - a^2)I^f_{F_b} \\ (a^2 - a)I^f_{F_b} \end{bmatrix}
$$

$$
\therefore I^f_{F_0} = 0; \quad I^f_{F_1} = -I^f_{F_2} = \frac{1}{3}(a - a^2)I^f_{F_b}.
$$

Because the zero-sequence current is zero, the zero-sequence voltage is also zero. The positive-sequence voltage can be written as:

$$
\begin{aligned}
V^f_{F_1} &= \frac{1}{3}\left(V^f_{F_a} + aV^f_{F_b} + a^2V^f_{F_c}\right) \\
&= \frac{1}{3}\left(V^f_{F_a} + aV^f_{F_b} + a^2(V^f_{F_b} - R_f I^f_{F_b})\right) \\
&= \frac{1}{3}\left(V^f_{F_a} + aV^f_{F_b} + a^2V^f_{F_b} - a^2 R_f I^f_{F_b}\right).
\end{aligned} \tag{2.20}
$$

The negative-sequence voltage can be written as:

$$
\begin{aligned}
V^f_{F_2} &= \frac{1}{3}\left(V^f_{F_a} + a^2V^f_{F_b} + aV^f_{F_c}\right) \\
&= \frac{1}{3}\left(V^f_{F_a} + a^2V^f_{F_b} + a(V^f_{F_b} - R_f I^f_{F_b})\right) \\
&= \frac{1}{3}\left(V^f_{F_a} + a^2V^f_{F_b} + aV^f_{F_b} - a R_f I^f_{F_b}\right).
\end{aligned} \tag{2.21}
$$

Substituting (2.20) in (2.21), we get:

$$
V^f_{F_2} = V^f_{F_1} + \frac{1}{3}R_f I^f_{F_b}(a^2 - a)
$$

$$
\therefore V^f_{F_1} - V^f_{F_2} = -\frac{1}{3}R_f I^f_{F_b}(a^2 - a) \tag{2.22}
$$

$$
= R_f I^f_{F_1} \quad \because I^f_{F_1} = \frac{1}{3}(a - a^2)I^f_{F_b}.
$$

From (2.19) and (2.22), we can draw two conclusions. First, during a line-to-line fault, only positive- and negative-sequence components are present. Second, the positive-sequence and negative-sequence networks are connected in parallel through R_f. This is shown in Figure 2.2(c). The fault point in the positive-sequence network (F_1) is connected to the fault point in the negative-sequence network (F_2) through a fault resistance of R_f. The reference bus (N_1) of the positive-sequence network is connected to the reference bus (N_2) of the negative-sequence network.

Equations (2.19) and (2.22) were used to determine how to interconnect the sequence networks. Those results were derived by assuming an ABC phase rotation and using phase A as the reference for a BC fault. The results would have been the same for a system with ACB phase rotation. However, if any other phase was used as the reference for the BC fault, the results would have been different and the connection of the sequence networks shown in Figure 2.2(c) would no longer be valid. Therefore, choosing the correct reference phase is important. For line-to-line faults, always use the unfaulted phase as the reference phase.

(d) Double Line-to-Ground Fault

Figure 2.4 shows a BC-G fault that has a fault resistance of R_g between the common connection of the faulted phases and ground. For simplicity, assume the system was unloaded at the time of the fault (unfaulted A-phase current is zero). The following relations are true at the fault point:

$$I^f_{F_a} = 0, \tag{2.23}$$

$$V^f_{F_b} = V^f_{F_c} = R_g \times (I^f_{F_b} + I^f_{F_c}). \tag{2.24}$$

Assuming an ABC phase rotation, the sequence currents at the fault point are:

$$
\begin{aligned}
\begin{bmatrix} I^f_{F_0} \\ I^f_{F_1} \\ I^f_{F_2} \end{bmatrix} &= \frac{1}{3} \begin{bmatrix} 1 & 1 & 1 \\ 1 & a & a^2 \\ 1 & a^2 & a \end{bmatrix} \times \begin{bmatrix} I^f_{F_a} \\ I^f_{F_b} \\ I^f_{F_c} \end{bmatrix} \\
&= \frac{1}{3} \begin{bmatrix} 1 & 1 & 1 \\ 1 & a & a^2 \\ 1 & a^2 & a \end{bmatrix} \times \begin{bmatrix} 0 \\ I^f_{F_b} \\ I^f_{F_c} \end{bmatrix} \\
&= \frac{1}{3} \begin{bmatrix} I^f_{F_b} + I^f_{F_c} \\ aI^f_{F_b} + a^2 I^f_{F_c} \\ a^2 I^f_{F_b} + aI^f_{F_c} \end{bmatrix},
\end{aligned} \tag{2.25}
$$

$$\therefore I^f_{F_0} + I^f_{F_1} + I^f_{F_2} = 0 \quad (\because 1 + a + a^2 = 0).$$

Figure 2.4 BC-G fault on an unloaded power system.

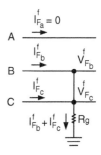

Assuming an ABC phase rotation, the sequence voltages at the fault point are:

$$
\begin{bmatrix} V_{F_0}^f \\ V_{F_1}^f \\ V_{F_2}^f \end{bmatrix} = \frac{1}{3} \begin{bmatrix} 1 & 1 & 1 \\ 1 & a & a^2 \\ 1 & a^2 & a \end{bmatrix} \times \begin{bmatrix} V_{F_a}^f \\ V_{F_b}^f \\ V_{F_c}^f \end{bmatrix}
$$

$$
= \frac{1}{3} \begin{bmatrix} 1 & 1 & 1 \\ 1 & a & a^2 \\ 1 & a^2 & a \end{bmatrix} \times \begin{bmatrix} V_{F_a}^f \\ V_{F_b}^f \\ V_{F_b}^f \end{bmatrix} \tag{2.26}
$$

$$
= \frac{1}{3} \begin{bmatrix} V_{F_a}^f + 2V_{F_b}^f \\ V_{F_a}^f + (a + a^2)V_{F_b}^f \\ V_{F_a}^f + (a + a^2)V_{F_b}^f \end{bmatrix},
$$

$$
\therefore V_{F_1}^f = V_{F_2}^f.
$$

Subtracting $V_{F_0}^f$ from $V_{F_2}^f$, we get:

$$
\begin{aligned}
V_{F_2}^f - V_{F_0}^f &= \frac{1}{3}\left(V_{F_a}^f + (a + a^2)V_{F_b}^f - V_{F_a}^f - 2V_{F_b}^f \right) \\
&= -V_{F_b}^f \\
&= -3R_g I_{F_0}^f \quad \because I_{F_b}^f + I_{F_c}^f = 3I_{F_0}^f, \\
\therefore V_{F_0}^f &= V_{F_2}^f + 3R_g I_{F_0}^f.
\end{aligned} \tag{2.27}
$$

Equations (2.25), (2.26), and (2.27) suggest that all three sequence components are present during a double line-to-ground fault. Furthermore, the three sequence networks must be connected in parallel and three times R_g must be placed between the negative-sequence and the zero-sequence networks as shown in Figure 2.2(d). The fault point in the positive-sequence network (F_1) is connected to the fault point in the negative-sequence network (F_2). F_2 is connected to the fault point in the zero-sequence network (F_0) through a fault resistance of $3R_f$. The reference bus of all the three networks are connected together. Fault resistance is multiplied by 3 to obtain the same voltage drop across the fault resistance with current $I_{F_0}^f$.

Note that the equations used to determine the interconnection of the sequence networks were derived assuming ABC phase rotation and using the A-phase as the reference for a BC-G fault. While the results would have been the same for a system with ACB phase rotation, the reference phase used is important. If any other phase were used as the reference for the AB-G fault, the results and hence the connection of the sequence networks would have been different. Therefore, reference phase selection is important. For double line-to-ground faults, use the unfaulted phase as the reference phase.

2.4 Sequence Impedances of Three-Phase Lines

This section explains how to calculate sequence impedance of three-phase lines and how to represent them in sequence networks. These impedances will be required as an input to impedance-based fault-locating algorithms in later chapters. Three methods are discussed. The first method calculates the sequence impedances by knowing the phase conductor type and spacing. The second method directly measures the sequence impedances while the third method estimates them from data captured during an unbalanced condition on the power system. Refer to [44] on how to calculate sequence impedances of other power system components.

Method 1: Calculating Sequence Impedances

This first step in this method is to calculate the self- and mutual impedances of the line. Carson's equations given below are typically used for this calculation [45, 46]. Applying these equations requires knowledge about the arrangement, geometric mean radius, and internal resistance of phase and neutral conductors [40].

$$z_{ii} = R_{Ci} + 0.00159f + j0.004657f \log_{10}\left(\frac{2160\sqrt{\frac{\rho}{f}}}{GMR_i}\right) \Omega/\text{mi}, \tag{2.28}$$

$$z_{ik} = 0.00159f + j0.004657f \log_{10}\left(\frac{2160\sqrt{\frac{\rho}{f}}}{d_{ik}}\right) \Omega/\text{mi}, \tag{2.29}$$

where z_{ii} is the self-impedance of conductor i in Ω/mi, z_{ik} is the mutual impedance between conductors i and k in Ω/mi, R_{Ci} is the internal resistance of conductor i in Ω/mi, f is the system frequency in Hz, GMR_i is the geometric mean radius of conductor i in ft, ρ is earth's resistivity in Ω/m, and d_{ik} is the distance between conductors i and k in ft. The second step is to build the line's primitive impedance matrix, Z_{prim}. Consider the three-phase, four-wire line shown in Figure 2.5. Here a, b, and c are the phase conductors and n is the neutral conductor. The voltage drop across the line can be written as:

$$\begin{bmatrix} V_{ag} \\ V_{bg} \\ V_{cg} \\ V_{ng} \end{bmatrix} = \begin{bmatrix} V'_{ag} \\ V'_{bg} \\ V'_{cg} \\ V'_{ng} \end{bmatrix} + \begin{bmatrix} Z_{aa} & Z_{ab} & Z_{ac} & Z_{an} \\ Z_{ba} & Z_{bb} & Z_{bc} & Z_{bn} \\ Z_{ca} & Z_{cb} & Z_{cc} & Z_{cn} \\ Z_{na} & Z_{nb} & Z_{nc} & Z_{nn} \end{bmatrix} \times \begin{bmatrix} I_a \\ I_b \\ I_c \\ I_n \end{bmatrix}, \tag{2.30}$$

where V_{ig} is the voltage between conductor i and the ground at one end of the line, V'_{ig} is the voltage between conductor i and the ground at the other end of the line, I_i is the current through conductor i, Z_{ii} is the self-impedance of conductor i, and Z_{ij} is the mutual impedance between conductors i and j. As you can see, the primitive impedance matrix has a size $N \times N$ where N is the total number of phase and neutral conductors. Denoting the phase and neutral conductors by subscripts p and n, respectively, we can partition the matrices in (2.30) and write it as:

$$\begin{bmatrix} [V_{pg}] \\ [V_{ng}] \end{bmatrix} = \begin{bmatrix} [V'_{pg}] \\ [V'_{ng}] \end{bmatrix} + \begin{bmatrix} [Z_{pp}] & [Z_{pn}] \\ [Z_{np}] & [Z_{nn}] \end{bmatrix} \times \begin{bmatrix} [I_p] \\ [I_n] \end{bmatrix}. \tag{2.31}$$

The third step is to remove the neutral conductor using Kron reduction. If the neutral is assumed to be perfectly grounded to the earth, V_{ng} and V'_{ng} can be set to zero. As a result,

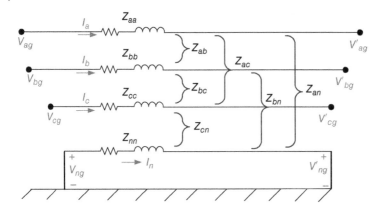

Figure 2.5 Kron reduction assumes a perfectly grounded neutral [45].

(2.31) can be rearranged and simplified to obtain a phase impedance matrix, Z_{abc}, of size $M \times M$, where M is the total number of phase conductors as:

$$[Z_{abc}] = [Z_{pp}] - [Z_{pn}][Z_{nn}]^{-1}[Z_{np}]. \tag{2.32}$$

The final step is transform the phase impedance to sequence impedance matrix as:

$$[Z_{seq}] = [A]^{-1} \times [Z_{abc}] \times [A] = \begin{bmatrix} Z_0 & 0 & 0 \\ 0 & Z_1 & 0 \\ 0 & 0 & Z_2 \end{bmatrix} \ \Omega/\text{mi}, \tag{2.33}$$

where Z_0 is the zero-sequence line impedance, Z_1 is the positive-sequence line impedance, and Z_2 is the negative-sequence line impedance. For three-phase lines, Z_1 is equal to Z_2, and Z_0 is higher than Z_1 (Z_0 is usually three times Z_1). Also note that matrix Z_{seq} in (2.33) has been shown to be a diagonal matrix. The off-diagonal terms are zero, indicating that there is no coupling between the three sequence networks. This is true if the conductor arrangement is symmetrical or if the line is transposed [43, 47]. However, when the line is not completely transposed and the conductor arrangement is not symmetrical, the off-diagonal numbers are usually small and can be ignored for the purposes of fault location.

Method 2: Measuring Sequence Impedances

Positive- and zero-sequence line impedance can also be measured by injecting current signals into the transmission line [48], [49]. This requires a line outage. At the remote end, all three phase and neutral connectors are shorted and connected to the ground. At the local end, a test set is used to inject a test current I_{test} into the following loops: a-g, b-g, c-g, a-b, b-c, c-a, and abc-g. Figure 2.6 shows the a-g, a-b, and abc-g test loops. The voltage measured in each test, V_{test}, is the voltage drop across the impedance of the loop being tested. Dividing V_{test} by I_{test} gives the impedance of the loop being tested. From these loop impedances, the positive- and zero-sequence line impedances can be calculated as:

$$Z_1 = \frac{\frac{1}{2}(Z_{ab} + Z_{bc} + Z_{ca})}{3}, \tag{2.34}$$

$$Z_0 = Z_1 + 3Z_e, \tag{2.35}$$

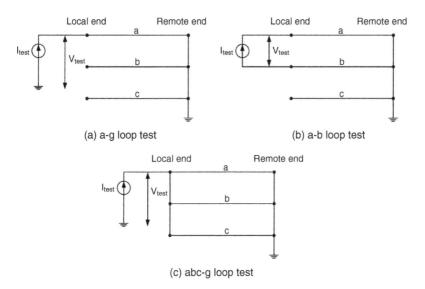

Figure 2.6 Injection test loops.

where $Z_e = \dfrac{Z_{ag} + Z_{bg} + Z_{cg}}{3} - Z_1,$ (2.36)

or $Z_e = Z_{abcg} - \dfrac{Z_1}{3},$ (2.37)

where Z_{ab}, Z_{bc}, and Z_{ca} are the impedances of a-b, b-c, and c-a loops, respectively, Z_{ag}, Z_{bg}, and Z_{cg} are the impedances of a-g, b-g, and c-g loops, respectively, and Z_{abcg} is the impedance of the abc-g loop.

Method 3: Estimating Sequence Impedances

The positive- and zero-sequence impedance of a line can be estimated using voltage and current phasors captured during an unbalanced condition on the power system such as a single-pole open condition, single line-to-ground fault, or double line-to-ground fault. In addition to the above listed scenarios, the positive-sequence line impedance can also be estimated during balanced conditions on the power system such as load or three-phase faults. But to estimate the zero-sequence impedance, the power system needs to be unbalanced with flow of zero-sequence current.

Single Pole Open
A two-terminal line with a single-pole open condition at terminal G is shown in Figure 2.7. A two-terminal line means there are sources at both ends of the line. Because pole A of the breaker at terminal G is open, the A-phase current ($I_{G_a}^{spo}$) at terminal G is zero while the other phases ($I_{G_b}^{spo}$ and $I_{G_c}^{spo}$) at that terminal carry load current. The single-pole open condition at terminal G is nothing but a series fault. At terminal H, all three poles of the breaker are closed. The following relations are true at terminal G assuming an ABC phase rotation:

$$\begin{aligned} I_{G_a}^{spo} &= 0, \\ I_{G_b}^{spo} &= aI_{G_c}^{spo}. \end{aligned}$$ (2.38)

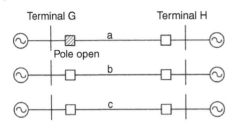

Terminal G   Terminal H

Pole open   a

                                  b

                                  c

Figure 2.7 Single-pole open condition on a two-terminal transmission line [121].

The symmetrical components during the single-pole open condition are:

$$I_{G_0}^{spo} = \frac{1}{3}\left(I_{G_a}^{spo} + I_{G_b}^{spo} + I_{G_c}^{spo}\right)$$

$$= \frac{1}{3}(1+a)I_{G_c}^{spo}$$

$$= -\frac{1}{3}a^2 I_{G_c}^{spo} \quad \because 1+a+a^2 = 0,$$

$$I_{G_1}^{spo} = \frac{1}{3}\left(I_{G_a}^{spo} + aI_{G_b}^{spo} + a^2 I_{G_c}^{spo}\right)$$

$$= \frac{2}{3}a^2 I_{G_c}^{spo},$$

$$I_{G_2}^{spo} = \frac{1}{3}\left(I_{G_a}^{spo} + a^2 I_{G_b}^{spo} + aI_{G_c}^{spo}\right)$$

$$= -\frac{1}{3}a^2 I_{G_c}^{spo},$$

$$\therefore I_{G_1}^{spo} = -\left(I_{G_0}^{spo} + I_{G_2}^{spo}\right).$$

(2.39)

Equation 2.39 indicates that all three sequence networks must be connected in parallel during a single-pole open condition as shown in Figure 2.8. The point of interconnection is at the point of the unbalance. Once the sequence networks are interconnected, the negative-sequence network can be used to estimate the positive-sequence line impedance [121]. The positive-sequence network is not used to minimize the effect of line-charging current. The positive-sequence line impedance is simply the negative-sequence voltage drop between terminal H and terminal G divided by the negative-sequence current flowing from terminal H to terminal G as shown below:

$$Z_1 = \frac{V_{H_2}^{spo} - V_{G_2}^{spo}}{I_{H_2}^{spo}}.$$

(2.40)

In a similar manner, the zero-sequence network can be used to estimate the zero-sequence line impedance as:

$$Z_0 = \frac{V_{H_0}^{spo} - V_{G_0}^{spo}}{I_{H_0}^{spo}}.$$

(2.41)

For accurate estimation of the sequence impedances, the method requires the line to carry significant load (load angle greater than five degrees) at the time of the single-pole open condition. The magnitude of load current determines the magnitude of negative- and zero-sequence current (their magnitude is one-third of the load current magnitude as

Figure 2.8 Interconnection of sequence networks during single-pole open on a two-terminal transmission line.

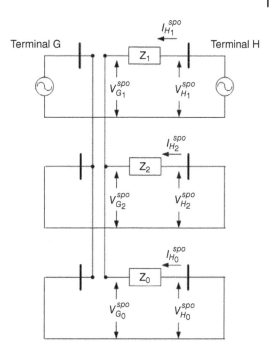

shown in (2.39)). When the magnitude of negative- and zero-sequence currents is small, any current or voltage measurement error gets amplified, thereby reducing the accuracy of the estimated sequence line impedance. Second, this method also requires line-side PTs to measure the negative- and zero-sequence voltage phasors on the line side of the open breaker ($V_{G_2}^{spo}$ and $V_{G_0}^{spo}$ in this example). Third, this method requires voltage phasors from both ends of the line. These phasors must be measured with respect to a common reference such as the Global Positioning System (GPS). The common reference allows measurements from different parts of the power system to be compared and used for mathematical calculations.

Ground Fault on Line

The sequence line impedance can also be estimated when there is a ground fault (single line-to-ground or double line-to-ground fault) on the line. Voltage and current phasors at one or both ends of the line during the fault are required. When using voltage and current phasors from both ends of the line, it is not required that the phasors from both line ends be synchronized. This approach is discussed in detail in Section 7.4.

2.5 Exercise Problems

■ Exercise 2.1

You are commissioning a newly installed relay in a distribution substation. The system has an ABC phase sequence and is operating under normal load conditions. The relay metering screen shows the phase currents measured by the relay and the calculated sequence currents. The currents are listed below.

$I_A = 100.52\angle 0°$ [A]

$I_B = 98.45\angle 119.80°$ [A]

$I_C = 103.78\angle -120.12°$ [A]

$I_1 = 1.65\angle 98.18°$ [A]

$I_2 = 100.92\angle -0.11°$ [A]

$I_0 = 1.45\angle -96.39°$ [A]

Do the sequence currents match what is expected for a balanced system? If no, what needs to be checked?

Solution:

In a balanced system, we expect to see mostly positive-sequence current and almost zero negative- and zero-sequence currents. The metering screen shows the relay calculating an unusually high negative-sequence current with almost zero positive-sequence current. Looking at the phase currents measured by the relay, the root cause becomes obvious. The relay is seeing ACB phase rotation even though it has been installed on a system with ABC phase rotation (relay phase rotation setting set to ABC). Most likely, the B-phase CT got wired to the C-phase input of the relay and the C-phase CT got wired to the B-phase input of the relay. Wiring errors such as the one in this example can lead to relay misoperations. This example demonstrates how sequence components can quickly catch wiring errors or relay setting errors and can help with commissioning relay systems.

■ Exercise 2.2

A three-phase distribution line has the pole configuration shown in Figure 2.9. The conductor type and its properties are listed in Table 2.1. Assume an earth resistivity (ρ) of 100 Ω-m and a system frequency (f) of 60 Hz.
  Calculate the sequence impedance matrix.

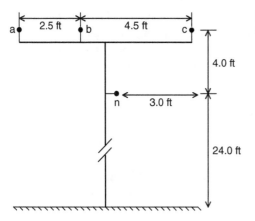

Figure 2.9 Spacing between conductors in exercise 2.2 [45].

Table 2.1 Conductor Data in Exercise 2.2

	Resistance (Ω/mi)	GMR (feet)
Phase 336,400 26/7 ACSR	0.306	0.0244
Shield 4/0 6/1 ACSR	0.592	0.00814

Solution:

Distances between phase conductors are:

$d_{ab} = 2.5$ ft,
$d_{bc} = 4.5$ ft,
$d_{ac} = 7$ ft.

Distance between the phase a conductor and the neutral conductor, d_{an}, is:

$$d_{an} = \sqrt{(2.5 + (4.5 - 3))^2 + 4^2} = 5.656 \;\; \text{ft}.$$

Distance between the phase b conductor and the neutral conductor, d_{bn}, is:

$$d_{bn} = \sqrt{(4.5 - 3)^2 + 4^2} = 4.272 \;\; \text{ft}.$$

Distance between the phase c conductor and the neutral conductor, d_{cn}, is:

$$d_{cn} = \sqrt{3^2 + 4^2} = 5 \;\; \text{ft}.$$

Using (2.28), we can calculate the self-impedance of the phase conductors ($Z_{aa} = Z_{bb} = Z_{cc}$) as:

$$Z_{ii} = R_{Ci} + 0.00159f + j0.004657f \log_{10}\left(\frac{2160\sqrt{\frac{\rho}{f}}}{GMR_i}\right)$$

$$= 0.306 + 0.0954 + j0.27942 \log_{10}\left(\frac{2160\sqrt{\frac{100}{60}}}{0.0244}\right)$$

$$= 0.4013 + j1.41333 \;\; \Omega/\text{mi}.$$

We can calculate the self-impedance of the neutral conductor (Z_{nn}) as:

$$Z_{ii} = R_{Ci} + 0.00159f + j0.004657f \log_{10}\left(\frac{2160\sqrt{\frac{\rho}{f}}}{GMR_i}\right)$$

$$= 0.592 + 0.0954 + j0.27942 \; \log_{10}\left(\frac{2160\sqrt{\frac{100}{60}}}{0.00814}\right)$$

$$= 0.6874 + j1.5465 \;\; \Omega/\text{mi}.$$

Using (2.29), we can calculate the mutual impedance between phase a and b conductors as:

$$Z_{ab} = 0.00159f + j0.004657f \log_{10} \left(\frac{2160\sqrt{\frac{\rho}{f}}}{d_{ab}} \right)$$

$$= 0.0954 + j0.27942 \log_{10} \left(\frac{2160\sqrt{\frac{100}{60}}}{2.5} \right)$$

$$= 0.0954 + j0.8515 \ \Omega/\text{mi}.$$

The mutual impedance between phase b and c conductors is:

$$Z_{bc} = 0.00159f + j0.004657f \log_{10} \left(\frac{2160\sqrt{\frac{\rho}{f}}}{d_{bc}} \right)$$

$$= 0.0954 + j0.27942 \log_{10} \left(\frac{2160\sqrt{\frac{100}{60}}}{4.5} \right)$$

$$= 0.0954 + j0.7802 \ \Omega/\text{mi}.$$

The mutual impedance between phase c and a conductors is:

$$Z_{ca} = 0.00159f + j0.004657f \log_{10} \left(\frac{2160\sqrt{\frac{\rho}{f}}}{d_{ca}} \right)$$

$$= 0.0954 + j0.27942 \log_{10} \left(\frac{2160\sqrt{\frac{100}{60}}}{7} \right)$$

$$= 0.0954 + j0.7266 \ \Omega/\text{mi}.$$

The mutual impedance between phase a and the neutral conductor is:

$$Z_{an} = 0.00159f + j0.004657f \log_{10} \left(\frac{2160\sqrt{\frac{\rho}{f}}}{d_{an}} \right)$$

$$= 0.0954 + j0.27942 \log_{10} \left(\frac{2160\sqrt{\frac{100}{60}}}{5.656} \right)$$

$$= 0.0954 + j0.7524 \ \Omega/\text{mi}.$$

The mutual impedance between phase b and the neutral conductor is:

$$Z_{bn} = 0.00159f + j0.004657f \log_{10}\left(\frac{2160\sqrt{\frac{\rho}{f}}}{d_{bn}}\right)$$

$$= 0.0954 + j0.27942 \log_{10}\left(\frac{2160\sqrt{\frac{100}{60}}}{4.272}\right)$$

$$= 0.0954 + j0.7865 \; \Omega/\text{mi}.$$

The mutual impedance between phase c and the neutral conductor is:

$$Z_{cn} = 0.00159f + j0.004657f \log_{10}\left(\frac{2160\sqrt{\frac{\rho}{f}}}{d_{cn}}\right)$$

$$= 0.0954 + j0.27942 \log_{10}\left(\frac{2160\sqrt{\frac{100}{60}}}{5}\right)$$

$$= 0.0954 + j0.7674 \; \Omega/\text{mi}.$$

Based on the above calculations, the primitive phase impedance matrix can be written as:

$$Z_{prim} = \left[\begin{array}{ccc|c} 0.4013 + j1.4133 & 0.0954 + j0.8515 & 0.0954 + j0.7266 & 0.0954 + j0.7524 \\ 0.0954 + j0.8515 & 0.4013 + j1.4133 & 0.0954 + j0.7802 & 0.0954 + j0.7865 \\ 0.0954 + j0.7266 & 0.0954 + j0.7802 & 0.4013 + j1.4133 & 0.0954 + j0.7674 \\ \hline 0.0954 + j0.7524 & 0.0954 + j0.7865 & 0.0954 + j0.7674 & 0.6873 + j1.5465 \end{array}\right] \Omega/\text{mi}$$

The 3×3 matrix in Z_{prim} is the phase impedance matrix. If the line had three wires instead of four wires, Z_{prim} would be equal to the phase impedance matrix. Since the line has 4 wires, Z_{prim} is a 4×4 matrix. Using (2.32) to perform Kron reduction and remove the neutral conductor, we get the below phase impedance matrix.

$$[Z_{abc}] = [Z_{pp}] - [Z_{pn}][Z_{nn}]^{-1}[Z_{np}]$$

$$= \begin{bmatrix} 0.4576 + j1.0780 & 0.1560 + j0.5017 & 0.1535 + j0.3849 \\ 0.1560 + j0.5017 & 0.4666 + j1.0482 & 0.1580 + j0.4236 \\ 0.1535 + j0.3849 & 0.1580 + j0.4236 & 0.4615 + j1.0651 \end{bmatrix} \Omega/\text{mi}.$$

The sequence impedance matrix is calculated from the phase impedance matrix using (2.33), as shown below.

$$[Z_{seq}] = [A]^{-1} \times [Z_{abc}] \times [A]$$

$$= \begin{bmatrix} 0.7736 + j1.9373 & 0.0256 + j0.0115 & -0.0321 + j0.0159 \\ -0.0321 + j0.0159 & 0.3061 + j0.6271 & -0.0723 - j0.0060 \\ 0.0256 + j0.0115 & 0.0723 - j0.0059 & 0.3061 + j0.6271 \end{bmatrix} \Omega/\text{mi}.$$

You can see that $Z_1 = Z_2 = 0.3061 + j0.6271$ Ω/mi, and $Z_0 = 0.7736 + j1.9373$ Ω/mi. The magnitude of Z_0 is three times the magnitude of Z_1. Because the arrangement of the phase conductors is not symmetrical, the off-diagonal terms are not zero. This indicates coupling between the three sequence networks. Transposing the transmission line makes the phase conductors balanced over the length of the line.

2.6 Summary

The key takeaways from this chapter as follows:

1) Symmetrical components is a mathematical tool that is used to analyze unbalanced faults on the power system. It decomposes an unbalanced set of three-phase phasors into three sets of balanced phasors: positive-, negative-, and zero-sequence components.
2) Positive-sequence components have the same magnitude, are 120 degrees apart, and have the same phase rotation as the system phase rotation. They are always present regardless of whether the power system is balanced or unbalanced.
3) Negative-sequence components have the same magnitude, are 120 degrees apart, and have a phase rotation that is opposite to the system phase rotation. They are present only when the power system becomes unbalanced.
4) Zero-sequence components have the same magnitude and are in phase with each other. They are present only when the power system becomes unbalanced and has a path to ground.
5) Knowledge of the system phase rotation and reference phase is required when calculating symmetrical components from phasors or when calculating phasors from symmetrical components.
6) The reference phase must be chosen based on the fault type as listed in Table 2.2. For single line-to-ground faults, use the faulted phase as the reference. For line-to-line or double line-to-ground faults, use the unfaulted or the healthy phase as the reference. For three-phase faults, use any one of the three phases as the reference.

Table 2.2 Reference Phase Selection

Fault Type	Reference Phase
A-G	A
B-G	B
C-G	C
BC or BC-G	A
CA or CA-G	B
AB or AB-G	C
ABC	A, B, or C

7) During a single line-to-ground fault, the positive-, the negative-, and the zero-sequence networks are connected in series.

8) During a line-to-line fault, the positive- and the negative-sequence networks are connected in parallel.

9) During a double line-to-ground fault, the positive-, the negative-, and the zero-sequence networks are connected in parallel.

10) Positive-sequence impedance of a transmission line is equal to the negative-sequence impedance of a transmission line. The zero-sequence impedance of a transmission line is greater than the positive-sequence impedance, typically by a factor of 3.

3

Fault Location on Transmission Lines

Fault location techniques for transmission lines can be classified into impedance-based and traveling-wave technologies. Impedance-based algorithms use fundamental frequency (60 Hz) voltage and current phasors recorded by digital relays, digital fault recorders, and other intelligent electronic devices (IEDs) during a fault to estimate the apparent impedance between the IED and location of the short-circuit fault. Given the line impedance in ohms per unit distance, the apparent impedance can be converted to a distance estimate. A number of impedance-based fault location algorithms have been developed for transmission networks. Those that use data captured by an IED at one end of the line are commonly referred to as one-ended impedance-based algorithms (also referred to as single-ended impedance-based algorithms), while those using data captured by IEDs at all ends of a multi-ended transmission line are referred to as multi-ended impedance-based algorithms.

Traveling-wave fault location algorithms, on the other hand, move away from 60 Hz and use high frequency traveling waves generated by the fault to determine fault location. Fault location is based on a very simple concept of physics, velocity equals distance over travel time. Similar to its counterpart, traveling-wave fault location techniques can also be classified into single- and double-ended algorithms.

In this chapter, we focus on presenting the underlying theory behind impedance-based and traveling-wave fault location algorithms. We highlight the motivation behind the development of each fault-locating algorithm, define the input data requirement of each algorithm, and identify the strengths and weaknesses of each algorithm. We then work through exercises based on field events to further solidify our concepts.

3.1 One-Ended Impedance-Based Fault Location Algorithms

One-ended impedance-based fault location algorithms estimate the location of a fault by looking into a transmission line from only one end [36]. Voltage and current phasors recorded by a digital relay at one end of the line during a fault are used to determine the apparent impedance between the relay and the fault. Given the line impedance in ohms, the per-unit distance to a fault can be easily obtained. The advantages of using one-ended algorithms are that they are straightforward to implement, yield reasonable location estimates, and require data from only one end of a line.

Fault Location on Transmission and Distribution Lines: Principles and Applications, First Edition.
Swagata Das, Surya Santoso, and Sundaravaradan N. Ananthan.
© 2022 John Wiley & Sons Ltd. Published 2022 by John Wiley & Sons Ltd.
Companion website: www.wiley.com/go/das/faultlocation

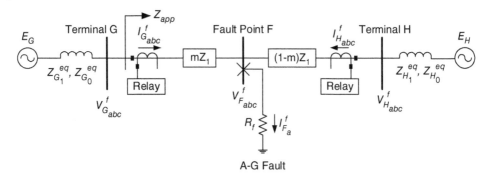

Figure 3.1 Two-terminal network.

We illustrate the principle of one-ended methods by using the two-terminal network shown in Figure 3.1. The overhead line between terminals G and H is homogeneous and has a positive- and zero-sequence impedance of Z_1 and Z_0 ohms, respectively. Recall from Chapter 2 that the negative-sequence impedance of the line is equal to the positive-sequence impedance of the line. The line is LL miles long. The network upstream from terminal G is represented by an ideal source, E_G, in series with an impedance that has positive- and zero-sequence values of $Z_{G_1}^{eq}$ and $Z_{G_0}^{eq}$ ohms, respectively. The network upstream from terminal H is also represented by an ideal source, E_H, in series with an impedance that has positive- and zero-sequence values of $Z_{H_1}^{eq}$ and $Z_{H_0}^{eq}$ ohms, respectively.

When the transmission line experiences an internal A-G fault with a resistance of R_f at m per unit distance from terminal G, both terminals contribute to the fault. Voltage and current phasors recorded by a digital relay at terminal G during the fault on all three phases are $V_{G_{abc}}^f$ volts and $I_{G_{abc}}^f$ amperes, respectively. Similarly, the voltage and current phasors recorded by a digital relay at terminal H during the fault on all three phases are $V_{H_{abc}}^f$ volts and $I_{H_{abc}}^f$ amperes, respectively. Voltage of all three phases at the fault point is $V_{F_{abc}}^f$ volts. The total phase current flowing into the fault is $I_{F_a}^f$ and is equal to the summation of $I_{G_a}^f$ and $I_{H_a}^f$. Because one-ended methods use measurements from only one terminal (terminal G or terminal H), $I_{F_a}^f$ is unknown since the contribution from the other terminal is not known.

Let us calculate the apparent impedance measured by the relay at terminal G during the fault. Figure 3.2 shows the interconnection of the positive-, negative-, and zero-sequence networks during the single line-to-ground fault. The positive-sequence voltage phasor at the fault point, $V_{F_1}^f$, can be calculated using the measurements from terminal G as:

$$V_{F_1}^f = V_{G_1}^f - mZ_1 \times I_{G_1}^f \quad [\text{V}], \tag{3.1}$$

where $V_{G_1}^f$ is the positive-sequence voltage phasor measured by the relay at terminal G during the fault in volts and $I_{G_1}^f$ is the positive-sequence current phasor recorded by the relay at terminal G during the fault in amperes. The negative-sequence voltage phasor at the fault point, $V_{F_2}^f$, can be written as:

$$V_{F_2}^f = V_{G_2}^f - mZ_1 \times I_{G_2}^f \quad [\text{V}], \tag{3.2}$$

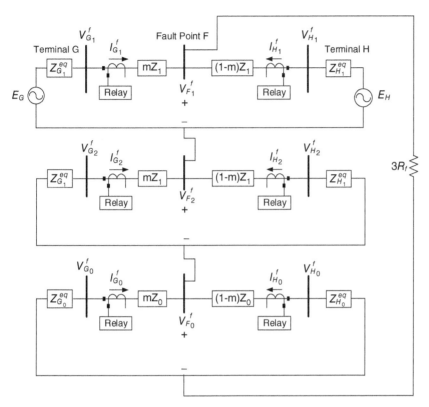

Figure 3.2 Interconnection of sequence networks during the single line-to-ground fault.

where $V^f_{G_2}$ is the negative-sequence voltage phasor measured by the relay at terminal G during the fault in volts and $I^f_{G_2}$ is the negative-sequence current phasor recorded by the relay at terminal G during the fault in amperes. The zero-sequence voltage phasor at the fault point, $V^f_{F_0}$, can be expressed as:

$$V^f_{F_0} = V^f_{G_0} - mZ_0 \times I^f_{G_0} \quad [V], \tag{3.3}$$

where $V^f_{G_0}$ is the zero-sequence voltage phasor measured by the relay at terminal G during the fault in volts and $I^f_{G_0}$ is the zero-sequence current phasor recorded by the relay at terminal G during the fault in amperes. Adding (3.1), (3.2), and (3.3), we get the voltage phasor of the faulted phase at the fault point, $V^f_{F_a}$, as:

$$\begin{aligned} V^f_{F_a} &= V^f_{F_0} + V^f_{F_1} + V^f_{F_2} \\ &= V^f_{G_a} - mZ_1 \left(I^f_{G_1} + I^f_{G_2} \right) - mZ_0 I^f_{G_0} \quad [V]. \end{aligned} \tag{3.4}$$

To further simplify, we add and subtract term $mZ_1 I^f_{G_0}$ from (3.4). We also express $V^f_{F_a}$ as the voltage drop across the fault resistance and get the following:

$$R_f I^f_{F_a} = V^f_{G_a} - mZ_1 \left(I^f_{G_1} + I^f_{G_2} + I^f_{G_0} \right) - mZ_0 I^f_{G_0} + mZ_1 I^f_{G_0}. \tag{3.5}$$

On rearranging the terms, we get:

$$V_{G_a}^f = mZ_1 I_{G_a}^f + mI_{G_0}^f (Z_0 - Z_1) + R_f I_{F_a}^f$$
$$= mZ_1 \left(I_{G_a}^f + k I_{G_0}^f \right) + R_f I_{F_a}^f \quad [\text{V}], \tag{3.6}$$

where k is the zero-sequence compensation factor and is defined as:

$$k = \frac{Z_0 - Z_1}{Z_1}. \tag{3.7}$$

The apparent impedance measured by the relay at terminal G during the A-G fault, Z_{app}, is:

$$Z_{app} = \frac{V_{G_a}^f}{I_{G_a}^f + k I_{G_0}^f} = mZ_1 + R_f \left(\frac{I_{F_a}^f}{I_{G_a}^f + k I_{G_0}^f} \right) \quad [\Omega]. \tag{3.8}$$

Generalizing the above equation for all fault types, we get:

$$Z_{app} = \frac{V_G}{I_G} = mZ_1 + R_f \left(\frac{I_F}{I_G} \right) \quad [\Omega], \tag{3.9}$$

where V_G and I_G are defined in Table 3.1 for all fault types. For single line-to-ground faults, V_G equals the faulted phase voltage and I_G equals the summation of the faulted phase current and k times the zero-sequence current. For line-to-line or double line-to-ground faults, V_G equals the line-to-line voltage between the faulted phases and I_G equals the line-to-line current between the faulted phases. For three-phase faults, any of the phase-to-phase loops can be used.

Equation 3.9 is the fundamental equation that governs all one-ended impedance-based fault location algorithms. Unfortunately, because measurements from only one end of the line are available, (3.9) has four unknowns: m, R_f, and the magnitude and phase angle of I_F. Several one-ended algorithms have been developed to eliminate R_f and I_F from the fault location computation and are discussed in detail below.

3.1.1 Simple Reactance Method

The simple reactance method capitalizes on the fact that fault resistance is resistive in nature and estimates the reactance to the fault [36]. The method assumes that currents I_F

Table 3.1 Definition of V_G and I_G for Different Fault Types

Fault Type	V_G	I_G
A-G	$V_{G_a}^f$	$I_{G_a}^f + k I_{G_0}^f$
B-G	$V_{G_b}^f$	$I_{G_b}^f + k I_{G_0}^f$
C-G	$V_{G_c}^f$	$I_{G_c}^f + k I_{G_0}^f$
AB, AB-G, ABC	$V_{G_a}^f - V_{G_b}^f$	$I_{G_a}^f - I_{G_b}^f$
BC, BC-G, ABC	$V_{G_b}^f - V_{G_c}^f$	$I_{G_b}^f - I_{G_c}^f$
CA, CA-G, ABC	$V_{G_c}^f - V_{G_a}^f$	$I_{G_c}^f - I_{G_a}^f$

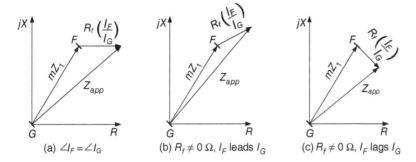

(a) $\angle I_F = \angle I_G$ (b) $R_f \neq 0\ \Omega$, I_F leads I_G (c) $R_f \neq 0\ \Omega$, I_F lags I_G

Figure 3.3 (a) When there is no load and the system is homogeneous, the simple reactance method is accurate with no reactance error even if the fault has resistance. (b) Reactance error is inductive due to a combination of load, a non-homogeneous system, and fault resistance. The fault appears farther away than it actually is. (c) Reactance error is capacitive due to a combination of load, a non-homogeneous system, and fault resistance. The fault appears closer than it actually is. [23].

and I_G are in phase with each other. As a result, the term $R_f\left(I_F/I_G\right)$ reduces to a real number as illustrated in Figure 3.3 (a). Considering only the imaginary components on both sides of (3.9), fault location from terminal G is given by:

$$m = \frac{\mathrm{imag}\left(\frac{V_G}{I_G}\right)}{\mathrm{imag}\left(Z_1\right)} \quad \text{[pu]}, \tag{3.10}$$

Multiplying the per unit value of m with LL gives the fault location in the same unit as LL.

While estimating the reactance to fault is an effective way to eliminate R_f and I_F from the fault location calculation, the accuracy of the simple reactance method is compromised when I_F and I_G are not in phase. This can occur due to load or a non-homogeneous system. A homogeneous system is one in which the source impedances have the same impedance angle as the line impedance. An example of such a system is shown in Figure 3.4. In reality, most transmission networks are non-homogeneous. Because the impedance angles are different, fault current contributed by each terminal have different phase angles. As a result, I_F, which is the summation of currents contributed by all terminals, has a different phase angle than I_G. The term $R_f\left(I_F/I_G\right)$ becomes a complex number and presents an additional reactance to the fault location calculation. Neglecting this reactance introduces an error in the location estimates that is commonly referred to as the reactance error.

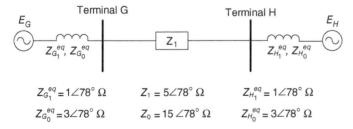

$$Z_{G_1}^{eq} = 1\angle 78°\ \Omega \qquad Z_1 = 5\angle 78°\ \Omega \qquad Z_{H_1}^{eq} = 1\angle 78°\ \Omega$$

$$Z_{G_0}^{eq} = 3\angle 78°\ \Omega \qquad Z_0 = 15\angle 78°\ \Omega \qquad Z_{H_0}^{eq} = 3\angle 78°\ \Omega$$

Figure 3.4 Example homogeneous system. The local and remote source impedances have the same impedance angle as the line impedance (78°).

When I_F leads I_G, for instance, the apparent impedance is greater than the actual impedance to fault. This inductive effect, shown in Figure 3.3 (b), causes the simple reactance method to overestimate the fault location. Or in other words, the fault appears to be farther away than it actually is. When I_F lags I_G, on the other hand, the apparent impedance is lower than the actual impedance to fault. This capacitive effect, shown in Figure 3.3 (c), causes the simple reactance method to underestimate the fault location. Or in other words, the fault appears to be closer than it actually is.

3.1.2 Takagi Method

The Takagi method [50] improves upon the performance of the simple reactance method by using the superposition principle and subtracting out the effect of load. The superposition principle states that a network during fault is the summation of a prefault network and a pure fault network. Figure 3.5 illustrates the principle of superposition when applied to the two-terminal network during a single line-to-ground fault on phase A. Notice that the pure fault network is driven by the positive-sequence voltage phasor at the fault point F before the fault, V_{F_1}. All other voltage sources are shorted. The pure fault current phasor, ΔI_G, can be derived as:

$$\Delta I_G = \Delta I_{G_0}^f + \Delta I_{G_1}^f + \Delta I_{G_2}^f, \tag{3.11}$$

where $\Delta I_{G_0}^f$, $\Delta I_{G_1}^f$, and $\Delta I_{G_2}^f$ are the zero-, positive-, and negative-sequence pure fault current phasors, respectively, measured by the relay at terminal G in amperes. Because the network is balanced before fault (no negative- or zero-sequence current), $\Delta I_{G_0}^f$ is equal to $I_{G_0}^f$. In a similar manner, $\Delta I_{G_2}^f$ is equal to $I_{G_2}^f$. Pure fault current $\Delta I_{G_1}^f$ is equal to the positive-sequence fault current phasor minus the prefault current phasor. Therefore, (3.11) can be written as:

$$\begin{aligned} \Delta I_G &= I_{G_0}^f + \left(I_{G_1}^f - I_{G_1} \right) + I_{G_2}^f \\ &= I_{G_a}^f - I_{G_a}. \end{aligned} \tag{3.12}$$

The above equation for ΔI_G is valid for an A-G fault. Table 3.2 defines ΔI_G for different fault types where $I_{G_{abc}}$ are the prefault current phasors recorded in all three phases at terminal G in amperes. Next, multiple both sides of (3.9) with the complex conjugate of ΔI_G and rearrange the terms to get:

$$V_G \times \Delta I_G^* = m Z_1 I_G \times \Delta I_G^* + R_f \left(I_F \times \Delta I_G^* \right). \tag{3.13}$$

Because the pure fault current phasor exists only during the fault, it is reasonable to assume that the phase angle of ΔI_G is the same as that of I_F in a homogeneous system. As a result, (3.13) reduces to:

$$V_G \times \Delta I_G^* = m Z_1 I_G \times \Delta I_G^* + R_f \times |I_F| \times |\Delta I_G|. \tag{3.14}$$

Saving only the imaginary components gets rid of the unknown fault resistance term and allows for fault location to be calculated from terminal G as:

$$m = \frac{\text{imag} \left(V_G \times \Delta I_G^* \right)}{\text{imag} \left(Z_1 \times I_G \times \Delta I_G^* \right)} \quad \text{[pu]}. \tag{3.15}$$

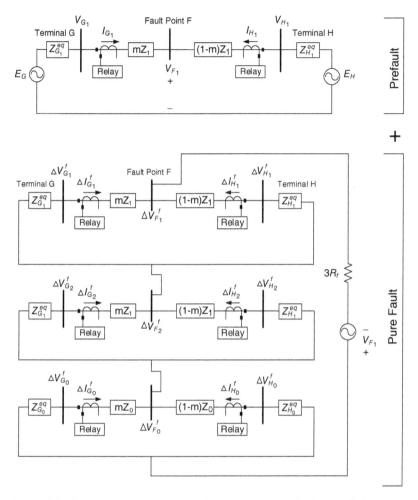

Figure 3.5 Superposition principle applied to the two-terminal transmission network during a single line-to-ground fault.

Table 3.2 Definition of ΔI_G for different fault types.

Fault Type	ΔI_G
A-G	$I^f_{G_a} - I_{G_a}$
B-G	$I^f_{G_b} - I_{G_b}$
C-G	$I^f_{G_c} - I_{G_c}$
AB, AB-G, ABC	$\left(I^f_{G_a} - I_{G_a}\right) - \left(I^f_{G_b} - I_{G_b}\right)$
BC, BC-G, ABC	$\left(I^f_{G_b} - I_{G_b}\right) - \left(I^f_{G_c} - I_{G_c}\right)$
CA, CA-G, ABC	$\left(I^f_{G_c} - I_{G_c}\right) - \left(I^f_{G_a} - I_{G_a}\right)$

Multiplying the per unit value of m with LL gives the fault location in the same unit as LL. Although the Takagi method makes clever use of pure fault current to minimize any reactance error due to system load, the success of this method relies on the network being homogeneous in nature. A non-homogeneous system will cause a reactance error that is proportional to the degree of non-homogeneity.

3.1.3 Modified Takagi Method

This method follows the same principle as the Takagi method to calculate fault location with one important difference. It uses the zero-sequence current instead of the pure fault current to account for load current during a single line-to-ground fault [51, 52]. This substitution is made possible since the zero-sequence current, similar to the pure fault current, exists only during a ground fault. The distance to fault from terminal G is computed as

$$m = \frac{\text{imag}\left(V_G \times 3I^{f*}_{G_0}\right)}{\text{imag}\left(Z_1 \times I_G \times 3I^{f*}_{G_0}\right)} \quad \text{[pu].} \tag{3.16}$$

Multiplying the per unit value of m with LL gives the fault location in the same unit as LL. The method can also compensate for a non-homogeneous system by applying an angle correction factor. The factor may be calculated by exercising the current division rule on the zero-sequence network shown in Figure 3.6 as:

$$\frac{I^f_{F_0}}{I^f_{G_0}} = \frac{Z^{eq}_{G_0} + Z_0 + Z^{eq}_{H_0}}{(1-m)Z_0 + Z^{eq}_{H_0}} = d_0 \angle \beta_0, \tag{3.17}$$

where d_0 is the current distribution factor for the zero-sequence network and angle β_0 represents the degree of non-homogeneity of the zero-sequence network. β_0 is zero in a homogeneous system (currents $I^f_{F_0}$ and $I^f_{G_0}$ in phase with each other) but has a finite value in a non-homogeneous system. It follows that applying an angle correction of $e^{-j\beta_0}$ to the fault location computation in (3.16) would force the system to be homogeneous and improve the accuracy of location estimates. However, to calculate β_0, the distance to fault, m, must be known. Therefore, locating single line-to-ground faults using the modified Takagi method requires three simple steps. First, calculate a preliminary estimate of m using (3.16). Second, use the preliminary estimate of m to calculate the angle correction factor β_0 in (3.17). Third,

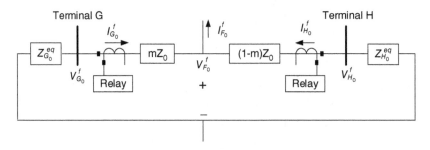

Figure 3.6 Zero-sequence network during a ground fault.

use the equation below to calculate fault location.

$$m = \frac{\text{imag}\left(V_G \times 3I_{G_0}^{f*} \times e^{-j\beta_0}\right)}{\text{imag}\left(Z_1 \times I_G \times 3I_{G_0}^{f*} \times e^{-j\beta_0}\right)} \quad \text{[pu]}. \tag{3.18}$$

The modified Takagi method with angle correction has a superior performance over the Takagi method as it addresses the two major sources of error, load and a non-homogeneous system. The success of the method, however, relies on accurately knowing the zero-sequence impedances of the local and remote sources. If the zero-sequence impedance of the local source is not available, it can be estimated as follows:

$$Z_{G_0}^{eq} = -\frac{V_{G_0}^f}{I_{G_0}^f} \quad [\Omega]. \tag{3.19}$$

The remote zero-sequence source impedance, on the other hand, must be known. Practically, this impedance is often not available. As a result, it is typical to apply the modified Takagi method without angle correction as given by (3.16).

Another version of the modified Takagi method uses the negative-sequence current instead of the zero-sequence current to account for load. Using the negative-sequence current has two benefits. First, negative-sequence current is present for all unbalanced faults. This allows the modified Takagi method to be used for locating any unbalanced fault. In contrast, using the zero-sequence current limits the use of the modified Takagi method to locating single line-to-ground faults only. Second, the negative-sequence network is more homogeneous than the zero-sequence network. (Transformer connections can make the zero-sequence network non-homogeneous.) As a result, fault location accuracy can improve when using the negative-sequence current over the zero-sequence current. Fault location can be obtained from the equation below.

$$m = \frac{\text{imag}\left(V_G \times I_{seq}^*\right)}{\text{imag}\left(Z_1 \times I_G \times I_{seq}^*\right)} \quad \text{[pu]}, \tag{3.20}$$

where I_{seq} depends on the fault type and is defined in Table 3.3.

3.1.4 Current Distribution Factor Method

This method uses current distribution factors to overcome any reactance error caused by fault resistance, load, and system non-homogeneity [53]. The distance to fault can be obtained by solving the quadratic equation below.

$$m^2 - k_1 m + k_2 - k_3 R_f = 0, \tag{3.21}$$

Table 3.3 Definition of I_{seq} for Different Fault Types

Fault Type	I_{seq}
single line-to-ground	$I_{G_2}^f$
line-to-line, double line-to-ground	$jI_{G_2}^f$

where constants k_1, k_2, and k_3 are complex functions of voltage, current, line impedance, and source impedances and are defined as follows:

$$k_1 = a + jb = 1 + \frac{Z_{H_1}^{eq}}{Z_1} + \left(\frac{V_G}{Z_1 \times I_G} \right),$$

$$k_2 = c + jd = \frac{V_G}{Z_1 \times I_G} \left(1 + \frac{Z_{H_1}^{eq}}{Z_1} \right),$$

$$k_3 = e + jf = \frac{\Delta I_G}{Z_1 \times I_G} \left(1 + \frac{Z_{H_1}^{eq} + Z_{G_1}^{eq}}{Z_1} \right).$$

Separating (3.21) into real and imaginary parts, the distance to fault m can be solved from the following quadratic equation:

$$m = \frac{\left(a - \frac{eb}{f} \right) \pm \sqrt{\left(a - \frac{eb}{f} \right)^2 - 4 \left(c - \frac{ed}{f} \right)}}{2} \quad \text{[pu]}, \tag{3.22}$$

where m can take two possible values. Usually one of the values will be positive and less than 1 per unit while the other value will either be positive but greater than 1 per unit or be negative. Since the fault location estimate must be positive and less that the total line length, the value of m that lies between 0 and 1 per unit should be chosen as the location estimate. Multiplying the chosen per unit value of m with LL gives the fault location in the same unit as LL.

If the local source impedance is not available, it can be calculated using (3.23) for unbalanced faults as:

$$Z_{G_1}^{eq} = Z_{G_2}^{eq} = -\frac{V_{G_2}^f}{I_{G_2}^f} \quad [\Omega]. \tag{3.23}$$

Use (3.24) to calculate the local source impedance for a three-phase fault.

$$Z_{G_1}^{eq} = -\frac{\Delta V_{G_1}^f}{\Delta I_{G_1}^f}$$

$$= -\frac{V_{G_1}^f - V_{G_1}}{I_{G_1}^f - I_{G_1}} \quad [\Omega], \tag{3.24}$$

where $\Delta V_{G_1}^f$ is the positive-sequence pure fault voltage phasor at terminal G in volts. The remote positive-sequence source impedance must be known.

3.2 Two-Ended Impedance-Based Fault Location Algorithms

Two-ended impedance-based fault location algorithms use waveform data captured at both ends of a two-terminal transmission line to estimate fault location. The fault-locating principle is similar to that of one-ended methods, i.e., using the voltage and current during a fault to estimate the apparent impedance between the monitoring location and the fault. Additional measurements from the remote end of the transmission line help eliminate any reactance error caused by fault resistance, load current, or system non-homogeneity. Depending on data availability, several two-ended methods have been developed in the literature. They are described below.

3.2.1 Synchronized Method

This method assumes that measurements from both ends of a transmission line are recorded at the same sampling rate and synchronized to a common time reference via the Global Positioning System (GPS) or other global navigation satellite system (GNSS). The GPS satellite system is owned by the United States government and provides geolocation and time information at no cost to a GPS receiver anywhere on earth. The receiver must have line of sight to at least four satellites in order to produce time signals with a 100 ns accuracy with respect to UTC (Coordinated Universal Time).

Any one of the three symmetrical components can be used for fault location computation. Using negative-sequence components is however, more advantageous since they are not affected by load current, zero-sequence mutual coupling, uncertainty in zero-sequence line impedance, or infeed from zero-sequence tapped loads [54, 55]. To illustrate the fault-locating principle, consider the negative-sequence network during an unbalanced fault as shown in Figure 3.7. The negative-sequence voltage at the fault point F, $V_{F_2}^f$, can be calculated from terminal G and H as

$$\text{Terminal G:} \quad V_{F_2}^f = V_{G_2}^f - mZ_1 I_{G_2}^f, \tag{3.25}$$

$$\text{Terminal H:} \quad V_{F_2}^f = V_{H_2}^f - (1-m)Z_1 I_{H_2}^f, \tag{3.26}$$

where $V_{G_2}^f$ and $I_{G_2}^f$ are the negative-sequence voltage and current phasors measured by the relay at terminal G during the fault, respectively, $V_{H_2}^f$ and $I_{H_2}^f$ are the negative-sequence voltage and current phasors measured by the relay at terminal H during the fault, respectively, and Z_1 is the positive-sequence line impedance. Since $V_{F_2}^f$ is equal when calculated from either terminal, equate (3.25) and (3.26) to solve for the distance to fault m as:

$$m = \frac{V_{G_2}^f - V_{H_2}^f + Z_1 I_{H_2}^f}{\left(I_{G_2}^f + I_{H_2}^f\right) Z_1} \quad \text{[pu]}. \tag{3.27}$$

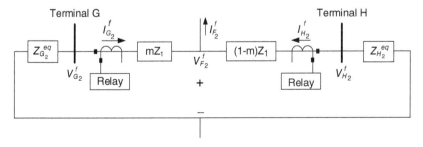

Figure 3.7 Negative-sequence network of a two-terminal line during an unbalanced fault.

Equation 3.27 is applicable for locating any unbalanced fault such as a single line-to-ground, line-to-line, or double line-to-ground fault. However, during a three-phase balanced fault, negative-sequence components do not exist. In such a case, the same fault-locating principle is applied to a positive-sequence network, and the distance to fault is computed as [56]

$$
m = \frac{V_{G_1}^f - V_{H_1}^f + Z_1 I_{H_1}^f}{\left(I_{G_1}^f + I_{H_1}^f \right) Z_1} \quad [\text{pu}],
\tag{3.28}
$$

where $V_{G_1}^f$ and $I_{G_1}^f$ are the positive-sequence voltage and current phasors measured by the relay at terminal G during the fault, respectively, and $V_{H_1}^f$ and $I_{H_1}^f$ are the positive-sequence voltage and current phasors measured by the relay at terminal H during the fault, respectively. This method does not require the knowledge of fault type. The presence or absence of negative-sequence components can be used to differentiate between an unbalanced and a balanced fault. Also note that multiplying the per unit value of m in (3.27) and (3.28) with LL gives the fault location in the same unit as LL.

3.2.2 Unsynchronized Method

Waveforms captured by IEDs at both ends of a transmission line may not be synchronized with each other. The IEDs can have different sampling rates or the GPS device may be absent or not functioning correctly. Therefore, to align the voltage and current measurements of terminal G with respect to terminal H, authors in [23] use a synchronizing operator $e^{j\delta}$ as,

$$
\text{Terminal G:} \quad V_{F_i}^f = V_{G_i}^f e^{j\delta} - m Z_1 I_{G_i}^f e^{j\delta},
\tag{3.29}
$$

$$
\text{Terminal H:} \quad V_{F_i}^f = V_{H_i}^f - (1 - m) Z_1 I_{H_i}^f,
\tag{3.30}
$$

where the subscript i refers to either the positive-sequence or negative-sequence component. Negative-sequence components are used to compute the location of an unbalanced fault while positive-sequence components are used to compute the location of a balanced three-phase fault. Equating (3.29) with (3.30), the synchronizing operator takes the form of

$$
e^{j\delta} = \frac{V_{H_i}^f - (1 - m) Z_1 I_{H_i}^f}{V_{G_i}^f - m Z_1 I_{G_i}^f}.
\tag{3.31}
$$

Now, $e^{j\delta}$ can be eliminated from the fault location computation by taking the absolute value on both sides of (3.31) as

$$
\left| e^{j\delta} \right| = 1 = \left| \frac{V_{H_i}^f - (1 - m) Z_1 I_{H_i}^f}{V_{G_i}^f - m Z_1 I_{G_i}^f} \right|.
\tag{3.32}
$$

Simplifying and rearranging the terms results in a quadratic equation. The unknown m can be solved by

$$
m = \frac{-B \pm \sqrt{B^2 - 4AC}}{2A} \quad [\text{pu}],
\tag{3.33}
$$

where the constants are defined as

$$
A = \left| Z_1 I_{G_i}^f \right|^2 - \left| Z_1 I_{H_i}^f \right|^2,
$$

$$B = -2 \times \text{Re} \left[V_{G_i}^f \left(Z_1 I_{G_i}^f \right)^* + \left(V_{H_i}^f - Z_1 I_{H_i}^f \right) \left(Z_1 I_{H_i}^f \right)^* \right],$$

$$C = \left| V_{G_i}^f \right|^2 - \left| V_{H_i}^f - Z_1 I_{H_i}^f \right|^2.$$

Solving the quadratic equation yields two values of m. One of the values is usually between 0 and 1 per unit and should be chosen as the location estimate. In the rare case that both values of m are between 0 and 1 per unit, use another symmetrical component to solve (3.33) and get a unique estimate for m. Multiplying the chosen per unit value of m with LL gives the fault location in the same unit as LL. This method does not require the knowledge of the fault type. The presence of negative-sequence quantities implies an unbalanced fault.

3.2.3 Unsynchronized Negative-Sequence Method

This method makes use of the negative-sequence network to calculate fault location. The negative-sequence fault voltage phasor from either terminal as [57]:

$$\text{Terminal G:} \quad V_{F_2}^f = - \left(Z_{G_1}^{eq} + mZ_1 \right) \times I_{G_2}^f, \tag{3.34}$$

$$\text{Terminal H:} \quad V_{F_2}^f = - \left(Z_{H_1}^{eq} + (1-m)Z_1 \right) \times I_{H_2}^f, \tag{3.35}$$

where $Z_{G_1}^{eq}$ and $Z_{H_1}^{eq}$ are the positive-sequence source impedance parameters behind terminals G and H, respectively. Equate (3.34) with (3.35) to eliminate $V_{F_2}^f$. Also, to avoid any alignment issues with data sets from both ends of the line, consider only the absolute values as

$$\left| I_{H_2}^f \right| = \left| \frac{\left(Z_{G_1}^{eq} + mZ_1 \right)}{\left(Z_{H_1}^{eq} + (1-m)Z_1 \right)} \times I_{G_2}^f \right|. \tag{3.36}$$

Squaring and rearranging the terms, we end up with a quadratic equation. Solve for m using (3.33) where the constants are defined as

$$A = \left| I_{H_2}^f \right|^2 \times (g^2 + h^2) - (c^2 + d^2),$$

$$B = -2 \times \left| I_{H_2}^f \right|^2 (eg + fh) - 2(ac + bd),$$

$$C = \left| I_{H_2}^f \right|^2 \times (e^2 + f^2) - (a^2 + b^2),$$

$$a + jb = I_{G_2}^f \times Z_{G_1}^{eq},$$

$$c + jd = Z_1 \times I_{G_2}^f,$$

$$e + jf = Z_{H_1}^{eq} + Z_1,$$

$$g + jh = Z_1.$$

Solve the quadratic equation to obtain two values of m. One of the values lies between 0 and 1 per unit and must be selected as the location estimate. Multiplying the chosen per unit value of m with LL gives the fault location in the same unit as LL. This method is applicable for locating unbalanced faults only. Knowledge of the fault type is not necessary. However,

Table 3.4 Definition of I_{pol} for Different Fault Types

Fault Type	I_{pol}
A-G, B-G, C-G	$I^f_{G_2} + I^f_{H_2}$
AB, AB-G, ABC	$(I^f_{G_a} + I^f_{H_a}) - (I^f_{G_b} + I^f_{H_b})$
BC, BC-G, ABC	$(I^f_{G_b} + I^f_{H_b}) - (I^f_{G_c} + I^f_{H_c})$
CA, CA-G, ABC	$(I^f_{G_c} + I^f_{H_c}) - (I^f_{G_a} + I^f_{H_a})$

the success of the method does depend on accurately knowing the source impedance parameters. If the local and remote source impedance is not known, (3.23) may be used to calculate the required parameters.

3.2.4 Synchronized Line Current Differential Method

This method is an extension of the modified Takagi method and takes advantage of the fact that measurements from the remote line terminal allow us to calculate the current at the fault point. Consider the negative-sequence network shown in Figure 3.7 during a single line-to-ground fault. Let us define a polarization current, I_{pol}, whose equation is given by:

$$I_{pol} = I^f_{G_2} + I^f_{H_2} = I^f_{F_2}.$$

(3.37)

This polarization current is nothing but the negative-sequence current at the fault point. This polarization current has the same phase angle as the phase fault current at the fault point. Therefore, multiplying both sides of (3.9) with the complex conjugate of I_{pol} and saving only the imaginary components, we can calculate the fault location as:

$$m = \frac{\text{imag}\left(V_G \times I^*_{pol}\right)}{\text{imag}\left(Z_1 \times I_G \times I^*_{pol}\right)} \quad \text{[pu]}.$$

(3.38)

The form taken by V_G and I_G, and I_{pol} for all fault types are defined in Table 3.1 and Table 3.4, respectively. Note that this method requires current data from both terminals of the line to be synchronized with each other and is best suited for implementation in a line current differential scheme. This is because in this scheme, the two relays are already set up to exchange synchronized current data with each other for line protection [58]. This fault location method is immune to load and non-homogeneous systems. However, unlike the other two-ended methods, it requires knowledge of the fault type and is affected by mutual coupling and inaccurate zero-sequence line impedance during a ground fault.

3.3 Three-Ended Impedance-Based Fault Location Algorithms

A three-terminal line is one that has three sources. Figure 3.8 shows an example of such a system. They are quite prevalent in utility transmission networks as they are

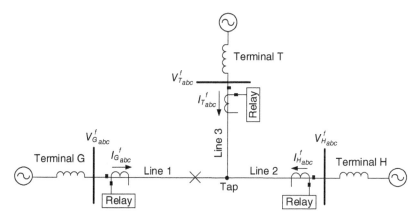

Figure 3.8 Three-terminal system.

a more cost- and time-effective solution to address immediate system requirements. They also provide increased voltage support, improve system performance, and offer operational flexibility [57]. Three-terminal lines may also be the only viable solution when right-of-way to construct new lines and stations are not available or when obtaining regulatory approvals is a concern [59]. On the downside, however, three-terminal lines increase the complexity of the network and pose a challenge to line protection and fault location. In this section, we adapt the two-ended fault location algorithms described in the previous section for application to a three-terminal line. The algorithms use information recorded by fault locators (relays in this example) at all three terminals to pinpoint the actual location of the fault. The approach is simple and consists of two steps. The first step identifies the line section with the fault. The second step reduces the three-terminal system into an equivalent two-terminal system and applies the two-ended fault location algorithms.

3.3.1 Synchronized Method

This method requires that all three relays be synchronized to a common GPS time clock. When an unbalanced fault occurs on Line 1, the negative-sequence network of the three-terminal line can be drawn as shown in Figure 3.9. It is evident from the figure that terminals T and H are operating in parallel to feed the fault on Line 1. As a result, the negative-sequence voltage phasor at the tap point ($V_{Tap_2}^f$) calculated from terminal H would be identical to that calculated from terminal T. The negative-sequence tap voltage phasor calculated from terminal G, on the other hand, would be quite different. If the fault was on Line 2 instead, the negative-sequence tap voltage phasor calculated from terminal G would be identical to that calculated from terminal T while the negative-sequence tap voltage phasor calculated from terminal H would be different. In a similar manner, if the fault was on Line 3, the negative-sequence tap voltage phasor calculated from terminal H would be identical to that calculated from terminal G while the negative-sequence tap voltage phasor calculated from terminal T would be different. We make use of these facts to identify the faulted line section.

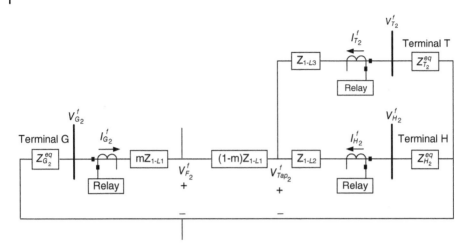

Figure 3.9 Negative-sequence network of the three-terminal system during an unbalanced fault.

First, calculate the negative-sequence tap voltage phasor from each terminal during the fault using the equations given below.

Terminal G: $\quad V^f_{Tap_{2@G}} = V^f_{G_2} - Z_{1-L1}I^f_{G_2},$ $\quad\quad\quad\quad\quad\quad\quad$ (3.39)

Terminal H: $\quad V^f_{Tap_{2@H}} = V^f_{H_2} - Z_{1-L2}I^f_{H_2},$ $\quad\quad\quad\quad\quad\quad\quad$ (3.40)

Terminal T: $\quad V^f_{Tap_{2@T}} = V^f_{T_2} - Z_{1-L3}I^f_{T_2}.$ $\quad\quad\quad\quad\quad\quad\quad$ (3.41)

Here $V^f_{G_2}$ and $I^f_{G_2}$ are the negative-sequence voltage and current phasors measured by the relay at terminal G during the fault, respectively, $V^f_{H_2}$ and $I^f_{H_2}$ are the negative-sequence voltage and current phasors measured by the relay at terminal H during the fault, respectively, $V^f_{T_2}$ and $I^f_{T_2}$ are the negative-sequence voltage and current phasors measured by the relay at terminal T during the fault, respectively, Z_{1-L1} is the positive-sequence impedance of Line 1, Z_{1-L2} is the positive-sequence impedance of Line 2, and Z_{1-L3} is the positive-sequence impedance of Line 3. The tap voltage phasor calculated using the faulted line section will be different than those calculated with the healthy line sections. In our example, you will find:

$$V^f_{Tap_{2@H}} = V^f_{Tap_{2@T}},$$
$$V^f_{Tap_{2@G}} \neq V^f_{Tap_{2@H}},$$
$$V^f_{Tap_{2@G}} \neq V^f_{Tap_{2@T}}.$$

This indicates that Line 1 is experiencing the fault. Once the faulted line has been identified, we refer to the two terminals operating in parallel to feed the fault as Remote Terminal 1 and Remote Terminal 2. We call the third terminal the Local Terminal.

Next, reduce the three-terminal system into an equivalent two-terminal system. This is illustrated in Figure 3.10 where $V^f_{loc_2}$ and $I^f_{loc_2}$ are the negative-sequence voltage and current phasors measured by the local relay, respectively, the positive-sequence impedance of the faulted line is Z_{1-line}, the positive-sequence source impedance behind the relay at the

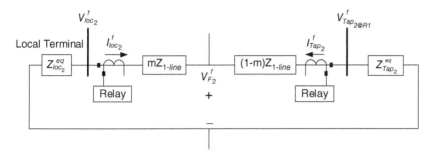

Figure 3.10 Negative-sequence network of the equivalent two-terminal system.

local terminal is $Z^{eq}_{loc_2}$, $V^f_{Tap_{2@R1}}$ is the negative-sequence tap voltage phasor calculated from Remote Terminal 1, $I^f_{Tap_2}$ is the summation of the negative-sequence currents contributed by the two remote terminals, and $Z^{eq}_{Tap_2}$ is the equivalent negative-sequence source impedance behind the tap point. In our example, the new notations are equivalent to:

$$V^f_{loc_2} = V^f_{G_2},$$

$$V^f_{Tap_{2@R1}} = V^f_{Tap_{2@H}},$$

$$I^f_{loc_2} = I^f_{G_2},$$

$$I^f_{Tap_2} = I^f_{H_2} + I^f_{T_2},$$

$$Z_{1-line} = Z_{1-L1},$$

$$Z^{eq}_{loc_2} = Z^{eq}_{G_2},$$

$$Z^{eq}_{Tap_2} = \frac{\left(Z_{1-L3} + Z^{eq}_{T_2}\right) \times \left(Z_{1-L2} + Z^{eq}_{H_2}\right)}{Z_{1-L2} + Z^{eq}_{H_2} + Z_{1-L3} + Z^{eq}_{T_2}},$$

where $Z^{eq}_{H_2}$ and $Z^{eq}_{T_2}$ are the source impedances of terminals H and T, respectively. We can then calculate the distance to an unbalanced fault as:

$$m = \frac{V^f_{loc_2} - V^f_{Tap_{2@R1}} + Z_{1-line} \times I^f_{Tap_2}}{\left(I^f_{loc_2} + I^f_{Tap_2}\right) Z_{1-line}}. \tag{3.42}$$

Negative-sequence quantities do not exist during a three-phase fault. Use the equation below to locate three-phase faults:

$$m = \frac{V^f_{loc_1} - V^f_{Tap_{1@R1}} + Z_{1-line} \times I^f_{Tap_1}}{\left(I^f_{loc_1} + I^f_{Tap_1}\right) Z_{1-line}}. \tag{3.43}$$

3.3.2 Unsynchronized Method

This method uses voltage and current phasors captured by the relays at all terminals during a fault to calculate the fault location. Relays need not be time aligned with each other. The first step is to identify the faulted line section. To do this, calculate the negative-sequence voltage phasor at the tap point from all three terminals using (3.39), (3.40), and (3.41). The magnitude of the negative-sequence tap voltage phasor calculated using the faulted line section will be different than those calculated using the healthy line sections. In our example, $|V^f_{Tap_{2@H}}|$ will equal $|V^f_{Tap_{2@T}}|$ while $|V^f_{Tap_{2@G}}|$ will have a different value.

The second step is to reduce the three-terminal system into the equivalent two-terminal system shown in Figure 3.10. Follow the same procedure as outlined in the previous method. The only difficulty lies in the fact that the negative-sequence fault current at the tap point cannot be obtained by simply adding the negative-sequence fault currents measured by the relays at the remote terminals. This is because the relays are not synchronized with each other. To overcome this difficulty, calculate the alignment angle of the relay at Remote Terminal 2 with respect to the relay at Remote Terminal 1 as:

$$\delta = \frac{\angle V^f_{Tap_{2@R1}}}{\angle V^f_{Tap_{2@R2}}}. \tag{3.44}$$

Phase-shifting the negative-sequence current measured at the Remote Terminal 2 with angle delta aligns the current measured at that terminal with that measured at Remote Terminal 1. In our example, we can calculate the negative-sequence current phasor at the tap point as:

$$I^f_{Tap_2} = I^f_{H_2} + I^f_{T_2} e^{j\delta}. \tag{3.45}$$

Next, use the quadratic formula in (3.33) to obtain a location estimate where the constants are defined as

$$A = \left| Z_{1-line} I^f_{loc_2} \right|^2 - \left| Z_{1-line} I^f_{Tap_2} \right|^2,$$

$$B = -2 \times \mathrm{Re} \left[V^f_{loc_2} \left(Z_{1-line} I^f_{loc_2} \right)^* + \left(V^f_{Tap_{2@R1}} - Z_{1-line} I^f_{Tap_2} \right) \left(Z_{1-line} I^f_{Tap_2} \right)^* \right],$$

$$C = \left| V^f_{loc_2} \right|^2 - \left| V^f_{Tap_{2@R1}} - Z_{1-line} I^f_{Tap_2} \right|^2.$$

3.3.3 Unsynchronized Negative-Sequence Method

This method [57] uses only the current phasors recorded at each terminal during the fault and source impedance data to pinpoint the location of a fault. The three relays need not be synchronized with each other. Similar to the previous methods, the first step is to identify whether Line 1, Line 2, or Line 3 experienced the fault. This is achieved by using the equations below to calculate the negative-sequence voltage phasor at the tap point from each of the three terminals:

$$\text{Terminal G:} \quad V^f_{Tap_{2@G}} = -\left(Z^{eq}_{G_2} + Z_{1-L1} \right) \times I^f_{G_2}, \tag{3.46}$$

$$\text{Terminal H:} \quad V^f_{Tap_{2@H}} = -\left(Z^{eq}_{H_2} + Z_{1-L2} \right) \times I^f_{H_2}, \tag{3.47}$$

$$\text{Terminal T:} \quad V^f_{Tap_{2@T}} = -\left(Z^{eq}_{T_2} + Z_{1-L3} \right) \times I^f_{T_2}. \tag{3.48}$$

Compare the magnitudes of the tap voltage. Tap voltage calculated using the faulted line will be different than those calculated using the healthy lines. In our example, $|V^f_{Tap_{2@H}}|$ will equal $|V^f_{Tap_{2@T}}|$ while $|V^f_{Tap_{2@G}}|$ will have a different value.

After identifying the faulted line section, reduce the three-terminal system into the equivalent two-terminal system shown in Figure 3.10 by following the same procedure outlined in the previous two methods. Note that the negative-sequence fault current at the tap point is the phasor addition of the negative-sequence fault currents measured by the relays at the two remote terminals. However, because the relays are not synchronized with each other, we can add the currents after performing an angle correction. Calculate the alignment angle of the relay at Remote Terminal 2 with respect to the relay at Remote Terminal 1 using (3.44). Phase-shift the negative-sequence current measured at Remote Terminal 2 with angle delta and add the result with the negative-sequence current measured at terminal 1 to calculate the negative-sequence current at the tap point using (3.45). Finally, use the quadratic formula in (3.33) to obtain a location estimate where the constants are defined as

$$a + jb = I^f_{loc_2} \times Z^{eq}_{loc_1},$$

$$c + jd = Z_{1-L1} \times I^f_{loc_2},$$

$$e + jf = Z^{eq}_{Tap_2} + Z_{1-L1},$$

$$g + jh = Z_{1-L1},$$

$$A = \left| I^f_{Tap_2} \right|^2 \times \left(g^2 + h^2 \right) - \left(c^2 + d^2 \right),$$

$$B = -2 \times \left| I^f_{Tap_2} \right|^2 (eg + fh) - 2 \left(ac + bd \right),$$

$$C = \left| I^f_{Tap_2} \right|^2 \times \left(e^2 + f^2 \right) - \left(a^2 + b^2 \right).$$

3.3.4 Synchronized Line Current Differential Method

This method [58] uses the fault voltage and current phasors recorded by the relay at the local end and synchronized fault current phasors from the other two ends of the line to identify the faulted line and calculate the fault location. It begins by assuming the fault to be on Line 1 and calculates the fault location from terminal G as:

$$m_G = \frac{\text{imag}\left(V_G \times I^*_{pol} \right)}{\text{imag}\left(Z_{1-L1} \times I_G \times I^*_{pol} \right)} \quad [\text{pu}], \tag{3.49}$$

where the polarization current is calculated from the local and remote currents. The fault is them assumed to be on Line 2, and the distance to fault from terminal H is given by:

$$m_H = \frac{\text{imag}\left(V_H \times I^*_{pol} \right)}{\text{imag}\left(Z_{1-L2} \times I_H \times I^*_{pol} \right)} \quad [\text{pu}]. \tag{3.50}$$

The fault is then assumed to be on Line 3, and the fault location is calculated from terminal T as:

$$m_T = \frac{\text{imag}\left(V_T \times I^*_{pol} \right)}{\text{imag}\left(Z_{1-L3} \times I_T \times I^*_{pol} \right)} \quad [\text{pu}]. \tag{3.51}$$

The line which yields a value of m less than 1 per unit is the one that experienced the fault and the corresponding value of m should be chosen as the fault location. In our example, m_H and m_T would have a value greater than 1 per unit while m_G would have a value less than 1 per unit. This suggests that the fault is on Line 1 and m_G is the location estimate. Note that when more than one value of m satisfies the above criterion, the implication is that the fault might be very close to the tap point.

3.4 Traveling-Wave Fault Location Algorithms

In this section, we are going to shift gears from the 60 Hz domain to the high frequency domain and discuss traveling waves. Traveling waves are generated by any disturbance on the power system that causes a step change in voltage such as a lightning strike, fault, or switching operation [60–62]. For example, suppose that a bolted fault occurs at the voltage peak on a 345 kV transmission line. The fault causes the peak voltage of the faulted phase to go from 282 kV to 0 V. This step change in voltage launches voltage traveling waves that have a magnitude of 282 kV in either direction from the fault. Associated with these voltage traveling waves are current traveling waves. The voltage and current traveling waves are related to each other by the characteristic impedance of the line, Z_C, as shown below.

$$I_{TW} = \frac{V_{TW}}{Z_C}. \tag{3.52}$$

The characteristic impedance depends on the inductance of the line (L) and the capacitance to ground (C) and can be calculated as $\sqrt{\frac{L}{C}}$. It is typically between 300 and 400 Ω for overhead lines and between 50 and 70 Ω for underground cables [63].

The current and voltage traveling waves travel with a velocity that is a percent of the speed of light. The percent is referred to as the line propagation velocity (v_p) which equals 98 percent on overhead lines and 55 percent on underground cables. When talking about traveling waves, it is typical to express their travel time in microseconds. For example, traveling waves would take 547.78 μs to travel 100 miles of overhead line.

When traveling waves reach a line terminal, they experience a change in the characteristic impedance since a terminal is a point of discontinuity. As a result, part of the wave gets reflected and the rest gets transmitted as shown in Figure 3.11. The magnitude of the reflected wave can be determined as:

$$i_R = \frac{Z_C - Z_T}{Z_T + Z_C} \times i_I = \rho_I \times i_I, \tag{3.53}$$

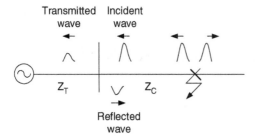

Transmitted wave   Incident wave

Z_T Z_C

Reflected wave

Figure 3.11 Traveling waves launched by the fault propagate in both directions from the fault. When the wave reaches a line terminal, part of the incident wave gets reflected while the rest gets transmitted.

where i_I is the incident wave, ρ_I is the reflection coefficient, Z_T is the terminating impedance, and i_R is the reflected wave. The magnitude of the transmitted wave can be determined as:

$$i_T = \frac{2 \times Z_C}{Z_T + Z_C} \times i_I = T_I \times i_I, \tag{3.54}$$

where i_T is the transmitted wave and T_I is the transmission coefficient. In a lossless line, the traveling waves would continue to bounce back and forth between terminals. Because transmission lines have losses, the traveling waves eventually attenuate and die down after multiple reflections.

Traveling waves contain valuable information about the fault. Algorithms that use them to locate faults are referred to as traveling-wave algorithms. There are several benefits to using traveling-wave fault location algorithms. First, these algorithms are highly accurate. Field data indicates an accuracy of one tower span (300 m or 1000 ft) [64, 65]. The accuracy is independent of line length, meaning that the one tower span accuracy is true for short lines as well as long lines. Second, because the algorithms do not use line impedance to calculate fault location, they are not affected by errors in the zero-sequence line impedance, by non-homogeneous lines, or by untransposed lines. Third, the accuracy is not affected by the existence of parallel lines or series capacitors. In fact, series capacitors appear as a short circuit to the high frequency traveling waves. Fourth, fault resistance, load, and fast clearing faults have no impact on fault location accuracy. Fifth and last, it is very easy to understand and calculate fault location with traveling-wave fault location algorithms. They are based on a very simple concept of physics, that is, velocity is equal to distance over time.

There are two types of traveling-wave fault location algorithms, single-ended and double-ended. The single-ended algorithm uses current traveling waves from one end of the line while double-ended algorithms use current traveling waves from both ends of the line. Fault locators that use single-ended traveling-wave algorithms are referred to as Type A fault locators while those that use double-ended traveling wave algorithms are referred to as Type D fault locators. Since current traveling waves can be measured by conventional CTs, no additional equipment is necessary to perform fault location [66]. The only requirement is that fault locators have a high-pass filter to reject the 60 Hz data and isolate the high frequency waves. When using traveling waves, the 60 Hz data become noise and the high frequency waves become useful data.

In this section, we discuss the traveling-wave fault location algorithms in more detail. We also discuss what affects the accuracy of traveling-wave fault location algorithms and how to overcome those error sources.

3.4.1 Single-Ended Traveling Wave Method

Refer to the one-line shown in Figure 3.12. An internal fault has occurred at a distance of m miles from Bus G on a line that is LL miles long. Below the one-line, a Bewley diagram shows the propagation of the waves along the transmission line and the multiple reflections and transmissions that occur when the waves reach a line terminal. In a Bewley diagram, the x-axis represents the line length and the y-axis represents time, where time increases in the downward direction. When a fault occurs at T_0, waves are launched at either direction from the fault. One wave arrives at Bus G at T_{G1}. Part of the wave reflects back to the fault,

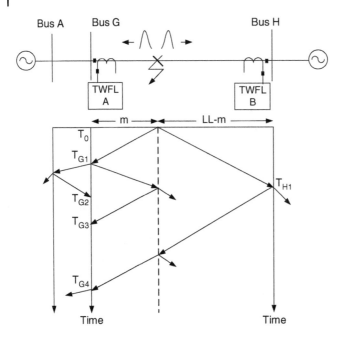

Figure 3.12 Bewley diagram during an internal fault.

gets reflected again, and comes back to Bus G at T_{G3}. Between T_{G3} and T_{G1}, the wave traveled twice the distance to the fault. Substituting this information in the velocity equation, we get:

$$2m = \text{velocity of traveling wave} \times \text{time of travel}$$

$$\therefore m = \frac{1}{2}\left(c \times v_p \times \left(T_{G3} - T_{G1}\right)\right). \tag{3.55}$$

Here c is the speed of light. Dividing throughout by LL, we can express the above equation as:

$$m = \frac{1}{2}\left(\frac{\left(T_{G3} - T_{G1}\right) \times LL}{T_{LL}}\right), \tag{3.56}$$

where T_{LL} is the traveling-wave line propagation time. This is the time taken by the wave to travel the one line length and can be expressed as:

$$T_{LL} = \frac{LL}{v_p \times c}. \tag{3.57}$$

The challenge with using the single-ended traveling-wave algorithm is that it is very difficult to identify the first reflection from the fault. If there is a short line behind the bus, it will generate reflections that are close to the reflection from the fault. For example, it is very difficult to determine whether the wave at T_{G2} or the wave at T_{G3} is the correct reflection from the fault in Figure 3.12. The knowledge that the reflection from the fault has the same polarity as the first wave can help with choosing the correct wave. References [67, 68] provide additional guidelines on how to sort through the waves and identify the correct wave.

3.4.2 Double-Ended Traveling-Wave Method

In the double-ended traveling-wave method, we have access to data captured by both the local and remote relay. This additional data from the remote relay simplifies the problem significantly and allows us to use the time stamps of when the local and remote relays saw the first wave to calculate fault location. Because we are interested in the first wave only, the problem of correctly identifying the reflection from the fault is taken out of the equation. Consider the same setup shown in Figure 3.12. When a fault occurs at time T_0 (unknown since we don't know when the fault occurred), traveling waves are launched at either direction from the fault. Suppose that the first wave reached the local end at time T_{G1}. The distance traveled by the wave is m and can be calculated as:

$$m = \text{velocity of traveling wave} \times \text{time of travel}$$
$$= c \times v_p \times \left(T_{G1} - T_0\right).$$

(3.58)

In a similar manner, the first wave from the fault reached the remote end at time T_{H1}. The distance traveled by the wave is $(LL - m)$ and can be calculated as:

$$LL - m = \text{velocity of traveling wave} \times \text{time of travel}$$
$$= c \times v_p \times \left(T_{H1} - T_0\right).$$

(3.59)

We end up with two equations and two unknowns: m and T_0. Writing the above equations in terms of T_0 and equating with each other, we can solve for m as:

$$m = \frac{v_p \times c \times \left(T_{G1} - T_{H1}\right) + LL}{2}.$$

(3.60)

Dividing the above equation by line length, we can express the above equation in terms of T_{LL} as:

$$m = \frac{LL}{2}\left(1 + \frac{T_{G1} - T_{H1}}{T_{LL}}\right).$$

(3.61)

This algorithm works with data from two relays. As a result, it requires both relays to be synchronized through GPS clocks or direct point-to-point fiber.

3.4.3 Error Sources

This section discusses the error sources that can affect the accuracy of traveling-wave methods.

(a) Traveling-Wave Line Propagation Time T_{LL}

This is an important input when calculating fault location using the traveling-wave methods. Accuracy analysis by [33] shows that a 1 microsecond error in the T_{LL} setting can result in a fault location error of 150 m on overhead lines. Some traveling-wave fault locators calculate T_{LL} by using a fixed value of v_p equal to 98.98 percent of the speed of light [69]. Other devices allow you to directly measure T_{LL} and enter the value as a user-defined setting. T_{LL} can be measured by performing a line energization test at the time of commissioning or by analyzing traveling-wave event records after an external or an internal fault [70]. These methods of measuring T_{LL} are outlined in detail below.

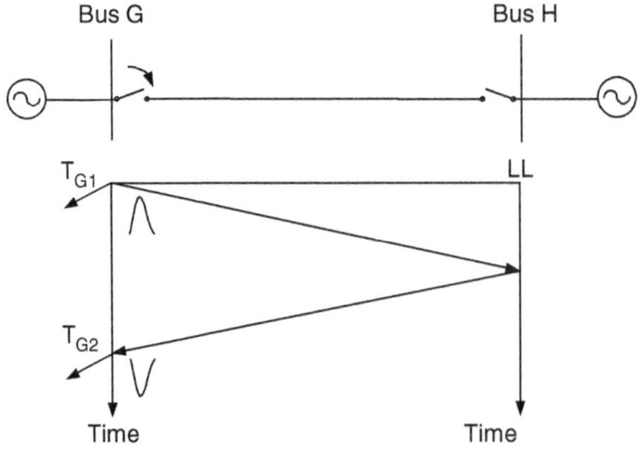

Figure 3.13 Determining T_{LL} from a line energization test.

Line Energization Test

During a line energization test, the breaker at the local end (Bus G in Figure 3.13) is closed to energize the line while keeping the remote breaker open. The step change in voltage when the breaker is closed, from zero to nominal, launches a current traveling wave at time T_{G1}. The current traveling wave travels to the remote breaker, gets reflected, and is reflected back to the local end with a polarity opposite to the first wave at time T_{G2}. Why does the reflected wave has an opposite polarity? To understand this, evaluate the reflection coefficient when the current wave arrives at Bus H. Because the remote breaker is open, impedance Z_T is infinite. As a result, 3.53 evaluates to −1 as shown below. This indicates that the incident wave will be completely reflected back with an opposite polarity.

$$\begin{aligned}\rho_I &= \frac{Z_C - Z_T}{Z_T + Z_C}\\ &= \frac{Z_C - \infty}{\infty + Z_C}\\ &= -1\end{aligned}$$

The traveling-wave fault locator (TWFL) at Bus G can be triggered to record waveform data during the line energization test. After downloading the data, identify the first wave and record the time stamp as T_{G1}. Next, identify the reflected wave by looking for a wave that has a polarity opposite to the first wave. Record the time stamp as T_{G2}. Between the two time stamps, the wave traveled twice the line length. Therefore, the T_{LL} can be easily calculated as:

$$T_{LL} = \frac{T_{G2} - T_{G1}}{2}. \tag{3.62}$$

The open remote breaker makes it easy to identify the reflection from the remote end.

Internal Fault

We can use the traveling wave data captured during an internal fault to calculate T_{LL}. Consider the internal fault shown in Figure 3.14. The fault at T_0 launches two waves, one going to Bus G and the other going to Bus H. The wave going toward Bus G reaches the bus at T_{G1}. Part of the wave reflects back to the fault, gets reflected again, and arrives back to Bus G

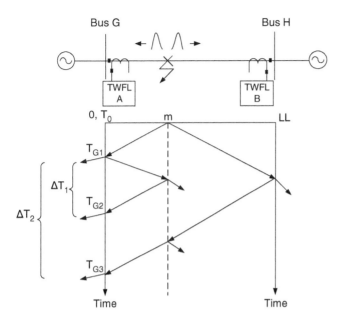

Figure 3.14 Determining T_{LL} using data captured during an internal fault.

at T_{G2}. The wave going toward Bus H reaches the bus, gets reflected, and arrives at Bus G at time T_{G3}. The distance traveled by the wave between T_0 and T_{G1} can be written as:

$$m = \text{velocity of traveling wave} \times \text{time of travel}$$
$$= c \times v_p \times (T_{G1} - T_0). \tag{3.63}$$

The distance traveled by the wave between T_0 and T_{G3} can be written as:

$$LL + (LL - m) = \text{velocity of traveling wave} \times \text{time of travel}$$
$$2LL - m = c \times v_p \times (T_{G3} - T_0) \tag{3.64}$$

Subtracting (3.64) from (3.63), we eliminate T_0 and end up with the equation below.

$$2(LL - m) = c \times v_p \times (T_{G3} - T_{G1})$$
$$= c \times v_p \times \Delta T_2. \tag{3.65}$$

The above equation has 2 unknowns: m and v_p. We can easily calculate m as:

$$m = \frac{c \times v_p \times (T_{G2} - T_{G1})}{2} = \frac{c \times v_p \times \Delta T_1}{2}. \tag{3.66}$$

Substituting (3.66) in (3.65), we can calculate T_{LL} as:

$$T_{LL} = \frac{\Delta T_1 + \Delta T_2}{2}. \tag{3.67}$$

This method of determining T_{LL} is more difficult than the line energization test. The challenge lies in correctly identifying the first reflection from the fault and the first reflection from the remote end among the reflections generated by adjacent lines or taps on the line.

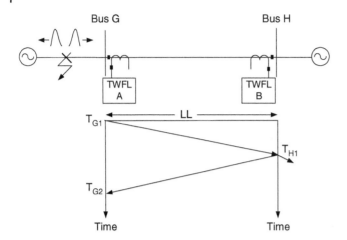

Figure 3.15 Determining T_{LL} using data captured during an external fault.

External Fault

We can also use traveling wave event records generated during an external fault to measure T_{LL}. Consider the external fault shown in Figure 3.15. One of the current traveling waves launched by the external fault goes back to the source while the other wave travels toward Bus G. At Bus G, part of the wave gets reflected, and the rest gets transmitted toward Bus H. This transmitted wave enters the polarity side of the CT. As a result, TWFL A measures the traveling wave current in the same direction as that on the line at T_{G1}. The wave travels down the line toward Bus H and eventually enters the non-polarity side of the CT. As a result, TWFL B sees the traveling wave current in the opposite direction to that on the line at T_{H1}. Once the wave reaches Bus H, part of it gets reflected back toward Bus G, and the rest gets transmitted through. The wave that got reflected toward Bus G is seen at T_{G2} by TWFL A.

When traveling wave event records from both ends are available, we are interested only in the first waves and can ignore subsequent reflections. Note down the polarity and the time stamp at which TWFL A saw the first wave (T_{G1}). Next, note down the time stamp at which TWFL B saw the first wave (T_{H1}) and confirm that the polarity of the wave is opposite to the polarity of the first wave seen by TWFL A. Calculate the T_{LL} as below:

$$T_{LL} = T_{H1} - T_{G1}. \tag{3.68}$$

If data from both ends is not available, we can use (3.62) to calculate T_{LL} by using data from one end only. It becomes challenging, however, to identify the reflection from the remote end. This problem was simplified in the line energization test by keeping the remote breaker open and by knowing that the reflected wave will have an opposite polarity to the first wave. Unfortunately, we don't have that benefit in this case.

(b) CT Cable Delay

When a traveling wave reaches the CT at a bus, it travels an additional distance equal to the CT secondary cable length before reaching the traveling-wave fault locator. Therefore, the wave arrival time recorded by the fault locator equals the travel time from the fault to the bus plus the travel time from the CT location to the fault locator. This additional time taken by the wave to reach the fault locator due to the CT cable is referred to as T_{CT} and is a source of fault location error. In fact, 0.1 microseconds of T_{CT} can cause a 15 m error in fault location on overhead lines [33]. Some traveling-wave fault locators allow users to measure the one-way CT cable length and enter that as a user-defined setting. The fault locator calculates T_{CT} (assuming a wave speed of 70 percent of the speed of light through CT secondary cables) and corrects the wave arrival time. For example, the double-ended traveling wave fault location equation given by (3.61) can be corrected for CT cable delay as:

$$m = \frac{LL}{2}\left(1 + \frac{(T_{G1} - T_{CT-G}) - (T_{H1} - T_{CT-H})}{T_{LL}}\right). \tag{3.69}$$

From the above equation, you can see that if the CT cable at Bus G is the same length as the CT cable at Bus H, T_{CT-G} will be equal to T_{CT-H} and they would cancel each other out. The CT cable delay only affects the double-ended traveling-wave fault location algorithm. In the single-ended traveling-wave fault location, this error source cancels out as shown below:

$$\begin{aligned} m &= \frac{(T_{G2} - T_{CT-G}) - (T_{G1} - T_{CT-G})}{2 \times T_{LL}} \times LL \\ &= \frac{(T_{G2} - T_{G1})}{2 \times T_{LL}} \times LL. \end{aligned} \tag{3.70}$$

(c) Line Length

There is a direct correlation between accuracy of line length and accuracy of traveling-wave fault location algorithms. A 1 percent error in line length can cause a 1 percent error in fault location [33]. Refer to 4.2.5 on the different definitions of line length. The line length setting that corresponds to the length of the transmission line conductor including sag is the most accurate definition.

(d) Hybrid Lines

Hybrid lines consist of both overhead line and underground cable sections. Underground cables while expensive improve aesthetics and allow transmission lines to cross through densely populated areas where it may be difficult to obtain a right-of-way permit. Traveling waves travel at almost 98 percent the speed of light through overhead sections but slow down considerably to 55 percent the speed of light when traveling through underground sections. This change in speed can cause hybrid lines to pose a challenge to those traveling-fault locators that assume propagation speed to remain the same for the entire line length.

Advanced traveling-wave fault locators [33] can locate faults on hybrid lines. They require users to enter the number of line sections and the line propagation time of each section. The fault locator then uses a three-step method to account for the change in propagation speed due to a hybrid line. The methodology is explained with the example hybrid line shown in Figure 3.16. The line is 15 miles long. The traveling wave line propagation time (T_{LL}) is 103.58 microseconds (54.78 microseconds to travel 10 miles of overhead line at 98 percent the speed of light and 48.8 microseconds to travel 5 miles of underground cable at 55 percent the speed of light). Now suppose that a fault occurs at 12 miles from Bus G at

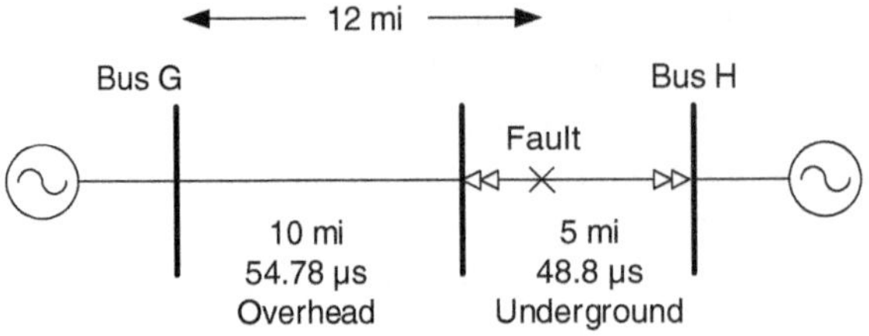

Figure 3.16 Hybrid line.

time t = 0 seconds, launching traveling waves in either direction. One wave reached Bus G at 12:39:44.663789 p.m. while the other wave reaches Bus H at 12:39:44.663744 p.m. The first step is to calculate fault location assuming a homogeneous line. In this example, using LL and T_{LL} in 3.61 we get:

$$m_{homogeneous} = \frac{LL}{2}\left(1 + \frac{T_{G1} - T_{H1}}{T_{LL}}\right)$$
$$= \frac{15}{2} \times \left(1 + \frac{45}{103.58}\right)$$
$$= 10.75 \text{ [mi]}.$$

This would be the fault location estimate from Bus G (fault location error of 1.25 miles) if one did not account for a hybrid line. To correct for a hybrid line, the second step is to calculate the propagation time corresponding to the distance estimate. In our example, the propagation time, T_{fault}, would be calculated as follows:

$$T_{fault} = \frac{103.58}{15} \times 10.75$$
$$= 74.23 \ \mu s.$$

The third step is to calculate the true fault location corresponding to this propagation time. In our example, the propagation time of the overhead section is 54.78 microseconds. So the fault must be beyond the overhead section and must have traveled an additional 19.45 microseconds on the underground section. Therefore, the true fault location is:

$$m_{true} = LL_{O/H} + \frac{\text{travel time to fault on U/G section}}{\text{total wave travel time on U/G section}} \times LL_{U/G}$$
$$= 10 + \frac{19.45}{48.8} \times 5$$
$$= 12 \text{ [mi]}.$$

(e) Fault at Voltage Zero
A fault that occurs at the zero-crossing of the voltage will not generate traveling waves. Since the voltage is already at zero volts at the time of fault inception, the fault cannot cause a sudden drop in voltage. No traveling waves means that the traveling wave fault locator cannot calculate a fault location. Traveling-wave fault locators such as the one shown in Figure 3.17 and Figure 7.18 have impedance-based fault location algorithms to back-up the traveling-wave methods for such a case. Fortunately, most insulation breakdowns that result in a fault occur at peak voltage conditions.

Figure 3.17 Protective relay equipped with traveling-wave and impedance-based fault location algorithms. (*Photo: Courtesy of Schweitzer Engineering Laboratories.*)

3.5 Exercise Problems

■ Exercise 3.1

A lightning strike caused a BC-G fault on a 161 kV transmission line during the middle of the day. The transmission line is radial from terminal G and has a total length of 12.58 miles before it reaches the next substation as shown in Figure 3.18. The positive- and zero-sequence line impedances are $Z_1 = 1.3790 + j9.3160\,\Omega$ and $Z_0 = 5.8841 + j25.8588\,\Omega$, respectively. The line impedances are in ohms primary. The fault was found to be 7.54 miles away from terminal G. A digital fault recorder (DFR) at the terminal captured the line-to-ground voltage and current waveforms during the event at 100 samples per cycle. They are shown in Figure 3.19. You can see that the fault lasted for a duration of three cycles before being cleared by a relay protecting the transmission line. The prefault current phasors during the twenty-sixth cycle are given below.

$$I_{G_{abc}}\ [\text{kA}]$$
$$\begin{bmatrix} 0.11\angle 157.43° \\ 0.11\angle 43.50° \\ 0.12\angle -80.25° \end{bmatrix}$$

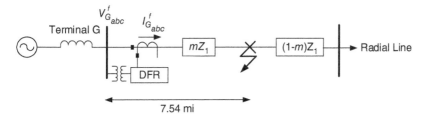

Figure 3.18 Exercise 3.1: One-line showing fault location.

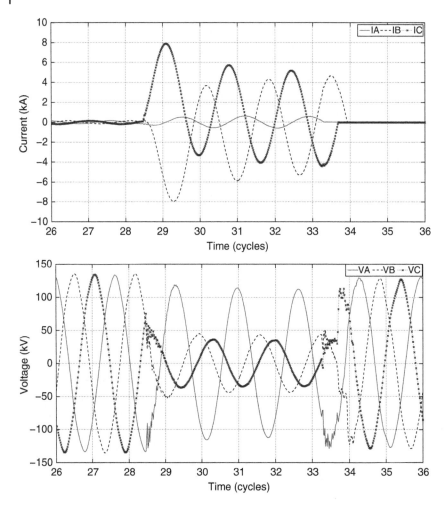

Figure 3.19 Exercise 3.1: Waveform data captured by the DFR at Terminal G during the lightning strike that caused a BC fault on the 161 kV transmission line.

The fault currents have a significant amount of DC offset. Before the DC offset could decay out completely, the fault was cleared from the line. All things considered, the thirty-first cycle appears to be the best cycle for calculating the voltage and current phasors during the fault and are given below:

$$I^f_{G_{abc}} \text{ [kA]} \qquad\qquad V^f_{G_{abc}} \text{ [kV]}$$

$$\begin{bmatrix} 0.42\angle 95.60° \\ 3.42\angle -33.15° \\ 3.35\angle -166.32° \end{bmatrix} \qquad\qquad \begin{bmatrix} 79.59\angle 153.03° \\ 29.87\angle 20.23° \\ 24.59\angle -62.06° \end{bmatrix}$$

Determine fault location using the following methods:

(a) Simple reactance method.
(b) Takagi method.
(c) Modified Takagi method.

Note that we cannot implement the current distribution factor method as this is a radial line.

Solution:

(a) Table 3.1 defines voltage V_G for a BC-G fault as the difference in voltage between the two faulted phases. The calculation is shown below.

$$V_G = V_{G_b}^f - V_{G_c}^f$$
$$= 29.87\angle 20.23° - 24.59\angle - 62.06°$$
$$= 36.05\angle 62.75° \quad [\text{kV}].$$

In a similar manner, I_G can be calculated as the difference between the currents flowing in the two faulted phases as shown below.

$$I_G = I_{G_b}^f - I_{G_c}^f$$
$$= 3.42\angle - 33.15° - 3.35\angle - 166.32°$$
$$= 6.21\angle - 10.00° \quad [\text{kA}].$$

Substituting V_G and I_G in (3.10), the distance to fault can be calculated as:

$$m = \frac{\text{imag}\left(\dfrac{V_G}{I_G}\right)}{\text{imag}(Z_1)}$$
$$= \frac{\text{imag}\left(\dfrac{36.05\angle 62.75°}{6.21\angle - 10.00°}\right)}{9.3160}$$
$$= 0.59 \quad [\text{pu}].$$

To obtain the fault location in units of miles, we multiply m with the line length and get 7.42 miles. The estimate is close to the actual fault location of 7.54 miles.

(b) The Takagi method uses the pure fault current to minimize any fault location error due to the combined effect of load and fault resistance. Per Table 3.2, the pure fault current for a BC-G fault can be calculated as:

$$\Delta I_G = \left(I_{G_b}^f - I_{G_b}\right) - \left(I_{G_c}^f - I_{G_c}\right)$$
$$= (3.42\angle - 33.15° - 0.11\angle 43.50°) - (3.35\angle - 166.32° - 0.12\angle - 80.25°)$$
$$= 6.19\angle - 11.84° \quad [\text{kA}].$$

Next, we calculate fault location as

$$m = \frac{\text{imag}\left(V_G \times \Delta I_G^*\right)}{\text{imag}\left(Z_1 \times I_G \times \Delta I_G^*\right)}$$
$$= \frac{\text{imag}\left(36.05\angle 62.75° \times 6.19\angle 11.84°\right)}{\text{imag}\left((1.3790 + j9.3160) \times 6.21\angle - 10.00° \times 6.19\angle 11.84°\right)}$$
$$= 0.60 \quad [\text{pu}].$$

Multiplying with the line length yields a location estimate of 7.55 miles.

(c) Knowledge of the negative-sequence current is necessary when applying the modified Takagi method. Because this is a BC-G fault, we choose the healthy phase, phase A, as the reference. By looking at the prefault voltage waveforms at the fifth cycle, the A-phase voltage peaks in the negative y-axis first, followed by B-phase and C-phase. From this, we can easily infer that the system has ABC phase rotation. Armed with this information, the sequence currents can be calculated as:

$$
\begin{bmatrix} I^f_{G_0} \\ I^f_{G_1} \\ I^f_{G_2} \end{bmatrix} = \frac{1}{3} \begin{bmatrix} 1 & 1 & 1 \\ 1 & a & a^2 \\ 1 & a^2 & a \end{bmatrix} \times \begin{bmatrix} I^f_{G_a} \\ I^f_{G_b} \\ I^f_{G_c} \end{bmatrix}
$$

$$
= \frac{1}{3} \begin{bmatrix} 1 & 1 & 1 \\ 1 & a & a^2 \\ 1 & a^2 & a \end{bmatrix} \times \begin{bmatrix} 0.42\angle 95.60° \\ 3.42\angle -33.15° \\ 3.35\angle -166.32° \end{bmatrix}
$$

$$
= \begin{bmatrix} 0.76\angle -100.57° \\ 2.38\angle 81.29° \\ 1.21\angle -102.43° \end{bmatrix} \quad [\text{kA}].
$$

We then calculate the fault location as follows:

$$
m = \frac{\text{imag}\left(V_G \times I^*_{seq}\right)}{\text{imag}\left(Z_1 \times I_G \times I^*_{seq}\right)}
$$

$$
= \frac{\text{imag}\left(36.05\angle 62.75° \times j1.21\angle 102.43°\right)}{\text{imag}\left((1.3790 + j9.3160) \times 6.21\angle -10.00° \times j1.21\angle 102.43°\right)}
$$

$$
= 0.60 \quad [\text{pu}].
$$

Multiplying with the line length gives a location estimate of 7.55 miles. This is a radial line, and hence there is no need to perform any angle correction.

In summary, all three methods were successful in pinpointing the fault location.

■ Exercise 3.2

An unfortunate bird bridged the gap between phase A and ground of a 161 kV two-terminal transmission line that connects terminal G and terminal H and created an A-phase-to-ground fault. The transmission line is 31.30 miles long and has a positive- and zero-sequence line impedance of $Z_1 = 5.0028 + j25.2730$ Ω and $Z_0 = 23.6659 + j79.3960$ Ω, respectively. The actual location of the fault was reported to be 30.86 miles from terminal G or 0.44 miles from terminal H as shown in Figure 3.20. Digital fault recorders at each terminal recorded the voltage and current waveforms during the fault at 96 samples per cycle. The waveforms are shown in Figure 3.21 and Figure 3.22.

The line is protected by distance relays at each terminal. Zone 1 in such relays is typically set equal to 80 percent of the line while Zone 2 is set equal to 120 percent of the line. At 0.44 miles, the fault was inside Zone 1 of the relay at terminal H and was cleared without any intentional time delay in about two and a half cycles. Once terminal H is open, the fault current contribution from terminal G increases at around the twelfth cycle. This is because the fault was outside Zone 1 but inside Zone 2 of the relay at terminal G. As a result, the fault

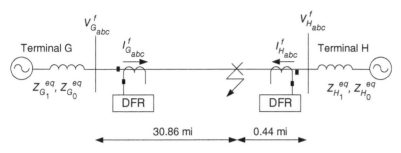

Figure 3.20 Exercise 3.2: One-line showing the reported location of the A-G fault.

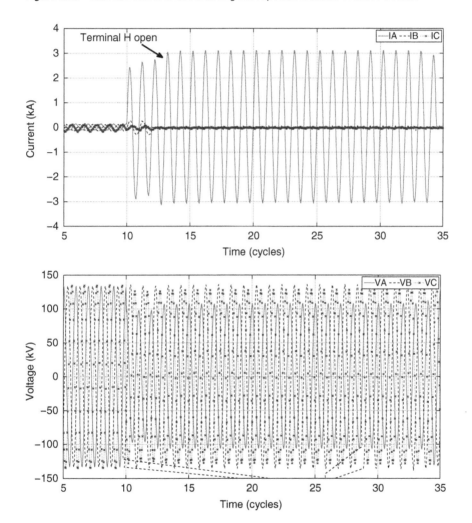

Figure 3.21 Exercise 3.2: DFR measurements at terminal G during the A-G fault that was caused by a bird coming in contact with a 161 kV transmission line.

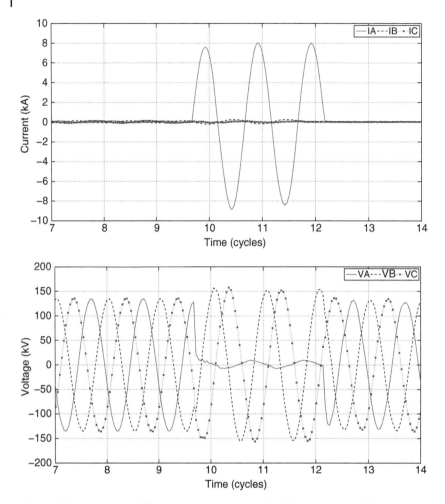

Figure 3.22 Exercise 2: DFR measurements at terminal H during the A-G fault that was caused by a bird coming in contact with a 161 kV transmission line.

stays on the system longer and is cleared after twenty-two and a half cycles by a distance relay at that terminal.

Voltage and current phasors recorded at terminal G during the twenty-eighth cycle are given below.

$$I^f_{G_{abc}} \text{ [kA]} \qquad\qquad V^f_{G_{abc}} \text{ [kV]}$$

$$\begin{bmatrix} 2.19\angle - 57.98° \\ 0.03\angle - 41.39° \\ 0.01\angle 179.70° \end{bmatrix} \qquad \begin{bmatrix} 75.45\angle 15.55° \\ 96.09\angle - 104.26° \\ 90.19\angle 132.54° \end{bmatrix}$$

During this cycle, only terminal G is contributing current to the fault with no remote infeed from terminal H. As a result, location estimates from one-ended methods are expected to be accurate. Determine the fault location using the modified Takagi method.

Solution:

Let us first calculate the zero-sequence current during the fault as:

$$3I_{G_0}^f = I_{G_a}^f + I_{G_b}^f + I_{G_c}^f$$

$$= 2.19\angle - 57.98° + 0.03\angle - 41.39° + 0.01\angle 179.70°$$

$$= 2.21\angle - 57.98° \quad [kA].$$

Next, let us calculate the zero-sequence compensation factor as:

$$k = \frac{Z_0 - Z_1}{Z_1}$$

$$= \frac{(23.6659 + j79.3960) - (5.0028 + j25.2730)}{5.0028 + j25.2730}$$

$$= 2.22\angle - 7.83°.$$

Current I_G during an A-G fault is:

$$I_G = I_{G_a}^f + k \times I_{G_0}^f$$

$$= 2.19\angle - 57.98° + \frac{2.22\angle - 7.83° \times 2.21\angle - 57.98°}{3}$$

$$= 3.82\angle - 61.33° \quad [kA].$$

Finally, calculate the fault location using (3.16) as:

$$m = \frac{\text{imag}\left(V_G \times 3I_{G_0}^{f*}\right)}{\text{imag}\left(Z_1 \times I_G \times 3I_{G_0}^{f*}\right)}$$

$$= \frac{\text{imag}\left(75.45\angle 15.55° \times 2.21\angle 57.98°\right)}{\text{imag}\left((5.0028 + j25.2730) \times 3.82\angle - 61.33° \times 2.21\angle 57.98°\right)}$$

$$= 0.76 \quad [pu].$$

After multiplying with the line length, we get a fault location estimate of 23.79 miles. Note that there is no need to correct for system non-homogeneity as we are calculating fault location using the phasors at the twenty-eighth cycle, when only terminal G is contributing to the fault with no infeed from terminal H. The fault location error is significant, 7.07 miles.

What caused the fault location error? The modified Takagi method is robust to fault resistance and load current. So they couldn't have been responsible for the error in fault location. Furthermore, non-homogeneous system does not come into play as a fault location error source in a radial line. During the cycle chosen for computing fault location, there was no DC offset in the current. The DFR was measuring line-to-ground voltages. This helps eliminate incorrect voltage and current phasors as possible sources of error. Most likely, the zero-sequence line impedance was inaccurate and gave rise to the error in fault location. This will become more evident in the next exercise. Chapter 7 describes an algorithm by which one might use fault event data to verify the accuracy of the zero-sequence line impedance.

■ **Exercise 3.3**

In this exercise, we are going to investigate whether two-ended methods can improve the accuracy of location estimates from the previous exercise. The two-ended methods will be applied to that part of the fault wherein both terminals were contributing to the fault. Voltage and current phasors at that time are given below.

$$
\begin{matrix}
I^f_{G_{abc}} \text{ [kA]} & V^f_{G_{abc}} \text{ [kV]} & I^f_{H_{abc}} \text{ [kA]} & V^f_{H_{abc}} \text{ [kV]} \\
\begin{bmatrix} 1.94\angle - 3.37° \\ 0.17\angle - 27.12° \\ 0.05\angle - 97.85° \end{bmatrix} &
\begin{bmatrix} 68.57\angle 68.41° \\ 95.74\angle - 50.75° \\ 88.36\angle - 176.66° \end{bmatrix} &
\begin{bmatrix} 5.84\angle 61.64° \\ 0.09\angle - 144.69° \\ 0.04\angle 154.49° \end{bmatrix} &
\begin{bmatrix} 5.41\angle 107.65° \\ 109.12\angle 7.47° \\ 108.95\angle - 91.38° \end{bmatrix}
\end{matrix}
$$

Determine the following:

(a) Negative-sequence current at terminal G during the fault.
(b) Negative-sequence voltage at terminal G during the fault.
(c) Negative-sequence current at terminal H during the fault.
(d) Negative-sequence voltage at terminal H during the fault.
(e) Negative-sequence source impedance at terminal G.
(f) Negative-sequence source impedance at terminal H.
(g) Fault location using unsynchronized negative-sequence method.
(h) Fault location using unsynchronized method.

Note that we can only apply the two-ended methods that use unsynchronized data from both terminals. This is because the DFRs at terminals G and H have different fault trigger times, and hence the waveforms that they record are not synchronized with each other.

Solution:

(a) The negative-sequence current at terminal G can be calculated by the equation below. We use the A-phase as the reference since this is an AG fault.

$$
\begin{bmatrix} I^f_{G_0} \\ I^f_{G_1} \\ I^f_{G_2} \end{bmatrix} = \frac{1}{3} \begin{bmatrix} 1 & 1 & 1 \\ 1 & a & a^2 \\ 1 & a^2 & a \end{bmatrix} \times \begin{bmatrix} I^f_{G_a} \\ I^f_{G_b} \\ I^f_{G_c} \end{bmatrix}
$$

$$
= \frac{1}{3} \begin{bmatrix} 1 & 1 & 1 \\ 1 & a & a^2 \\ 1 & a^2 & a \end{bmatrix} \times \begin{bmatrix} 1.94\angle - 3.37° \\ 0.17\angle - 27.12° \\ 0.05\angle - 97.85° \end{bmatrix}
$$

$$
= \begin{bmatrix} 0.70\angle - 6.61° \\ 0.63\angle 2.62° \\ 0.62\angle - 5.82° \end{bmatrix} \text{ [kA].}
$$

(b) The negative-sequence voltage at terminal G can be calculated as:

$$
\begin{bmatrix} V^f_{G_0} \\ V^f_{G_1} \\ V^f_{G_2} \end{bmatrix} = \frac{1}{3} \begin{bmatrix} 1 & 1 & 1 \\ 1 & a & a^2 \\ 1 & a^2 & a \end{bmatrix} \times \begin{bmatrix} V^f_{G_a} \\ V^f_{G_b} \\ V^f_{G_c} \end{bmatrix}
$$

$$= \frac{1}{3} \begin{bmatrix} 1 & 1 & 1 \\ 1 & a & a^2 \\ 1 & a^2 & a \end{bmatrix} \times \begin{bmatrix} 68.57\angle 68.41° \\ 95.74\angle -50.75° \\ 88.36\angle -176.66° \end{bmatrix}$$

$$= \begin{bmatrix} 5.24\angle -98.80° \\ 84.13\angle 66.96° \\ 10.94\angle -129.13° \end{bmatrix} \quad [\text{kV}].$$

(c) The negative-sequence current at terminal H can be calculated as:

$$\begin{bmatrix} I^f_{H_0} \\ I^f_{H_1} \\ I^f_{H_2} \end{bmatrix} = \frac{1}{3} \begin{bmatrix} 1 & 1 & 1 \\ 1 & a & a^2 \\ 1 & a^2 & a \end{bmatrix} \times \begin{bmatrix} I^f_{H_a} \\ I^f_{H_b} \\ I^f_{H_c} \end{bmatrix}$$

$$= \frac{1}{3} \begin{bmatrix} 1 & 1 & 1 \\ 1 & a & a^2 \\ 1 & a^2 & a \end{bmatrix} \times \begin{bmatrix} 5.84\angle 61.64° \\ 0.09\angle -144.69° \\ 0.04\angle 154.49° \end{bmatrix}$$

$$= \begin{bmatrix} 1.92\angle 62.44° \\ 1.96\angle 60.59° \\ 1.96\angle 61.92° \end{bmatrix} \quad [\text{kA}].$$

(d) The negative-sequence voltage at terminal H can be calculated as:

$$\begin{bmatrix} V^f_{H_0} \\ V^f_{H_1} \\ V^f_{H_2} \end{bmatrix} = \frac{1}{3} \begin{bmatrix} 1 & 1 & 1 \\ 1 & a & a^2 \\ 1 & a^2 & a \end{bmatrix} \times \begin{bmatrix} V^f_{H_a} \\ V^f_{H_b} \\ V^f_{H_c} \end{bmatrix}$$

$$= \frac{1}{3} \begin{bmatrix} 1 & 1 & 1 \\ 1 & a & a^2 \\ 1 & a^2 & a \end{bmatrix} \times \begin{bmatrix} 5.41\angle 107.65° \\ 109.12\angle 7.47° \\ 108.95\angle -91.38° \end{bmatrix}$$

$$= \begin{bmatrix} 45.74\angle -40.76° \\ 73.02\angle 137.32° \\ 22.64\angle -39.78° \end{bmatrix} \quad [\text{kV}].$$

(e) The negative-sequence source impedance at terminal G can be calculated using (3.23) as:

$$Z^{eq}_{G_2} = -\frac{V^f_{G_2}}{I^f_{G_2}}$$

$$= -\frac{10.94\angle -129.13°}{0.62\angle -5.82°}$$

$$= 9.7392 + j14.8215 \quad [\Omega].$$

(f) In a similar manner, the negative-sequence source impedance at terminal H can be calculated as:

$$Z^{eq}_{H_2} = -\frac{V^f_{H_2}}{I^f_{H_2}}$$

$$= -\frac{22.64\angle - 39.78°}{1.96\angle 61.92°}$$

$$= 2.3404 + j11.3064 \quad [\Omega].$$

(g) Let us first calculate constants a and b defined in Section 3.2.3 as:

$$a + jb = I_{G_2}^{f} \times Z_{G_2}^{eq}$$

$$= 0.62\angle - 5.82° \times (9.7392 + j14.8215)$$

$$= 6.9002 + j8.4828.$$

Constants c and d can be calculated as:

$$c + jd = Z_1 \times I_{G_2}^{f}$$

$$= (5.0028 + j25.2730) \times 0.62\angle - 5.82°$$

$$= 4.6481 + j15.1897.$$

Constants e and f can be calculated as:

$$e + jf = Z_{H_2}^{eq} + Z_1$$

$$= (2.3404 + j11.3064) + (5.0028 + j25.2730)$$

$$= 7.3432 + j36.5794.$$

Constants g and h can be calculated as:

$$g + jh = Z_1$$

$$= 5.0028 + j25.2730.$$

Constant A can be calculated as:

$$A = \left| I_{H_2}^{f} \right|^2 \times (g^2 + h^2) - (c^2 + d^2)$$

$$= 1.96^2 \times (5.0028^2 + 25.2730^2) - (4.6481^2 + 15.1897^2)$$

$$= 2298.70.$$

Constant B can be calculated as:

$$B = -2 \times \left| I_{H_2}^{f} \right|^2 (eg + fh) - 2(ac + bd)$$

$$= -2 \times 1.96^2 (7.3432 \times 5.0028 + 36.5794 \times 25.2730)$$

$$- 2(6.9002 \times 4.6481 + 8.4828 \times 15.1897)$$

$$= -7710.40.$$

Constant C can be calculated as:

$$C = \left| I_{H_2}^{f} \right|^2 \times (e^2 + f^2) - (a^2 + b^2)$$

$$= 1.96^2 \times (7.3432^2 + 36.5794^2) - (6.9002^2 + 8.4828^2)$$

$$= 5230.30.$$

Substituting A, B, and C in (3.33), we solve for the distance to fault:

$$m = \frac{-B \pm \sqrt{B^2 - 4AC}}{2A}$$

$$= \frac{7710.40 \pm \sqrt{(-7710.40)^2 - 4 \times 2298.70 \times 5230.30}}{2 \times 2298.70}$$

$$= 0.94, 2.41 \quad [\text{pu}].$$

Because $m = 0.94$ per unit satisfies the criteria of being positive and less than 1 per unit, it is deemed as the location estimate. Multiplying m with the line length yields a fault location estimate of 29.42 miles.

(h) First, we calculate constant A defined in Section 3.2.2 as:

$$A = \left| Z_1 I_{G_2}^f \right|^2 - \left| Z_1 I_{H_2}^f \right|^2$$

$$= \left| (5.0028 + j25.2730) \, 0.62\angle - 5.82° \right|^2 - \left| (5.0028 + j25.2730) \, 1.96\angle 61.92° \right|^2$$

$$= -2298.70.$$

Next, constant B is calculated as:

$$B = -2 \times \text{Re} \left[V_{G_2}^f \left(Z_1 I_{G_2}^f \right)^* + \left(V_{H_2}^f - Z_1 I_{H_2}^f \right) \left(Z_1 I_{H_2}^f \right)^* \right]$$

$$= -2 \times \text{Re} \left[10.94\angle - 129.13° \times \left((5.0028 + j25.2730) \times 0.62\angle - 5.82° \right)^* \right.$$

$$+ \left(22.64\angle - 39.78° - (5.0028 + j25.2730) \times 1.96\angle 61.92° \right)$$

$$\left. \times \left((5.0028 + j25.2730) \times 1.96\angle 61.92° \right)^* \right]$$

$$= 7710.40.$$

Finally, constant C is calculated as:

$$C = \left| V_{G_2}^f \right|^2 - \left| V_{H_2}^f - Z_1 I_{H_2}^f \right|^2$$

$$= 10.94^2 - \left| 22.64\angle - 39.78° - (5.0028 + j25.2730) \, 1.96\angle 61.92° \right|^2$$

$$= -5230.30.$$

Substituting A, B, and C in (3.33), we solved for the distance to fault:

$$m = \frac{-B \pm \sqrt{B^2 - 4AC}}{2A}$$

$$= \frac{-7710.40 \pm \sqrt{7710.40^2 - 4 \times (-2298.70) \times (-5230.30)}}{2 \times (-2298.70)}$$

$$= 0.94, 2.41 \quad [\text{pu}].$$

We choose $m = 0.94$ per unit, or 29.42 miles, as the location estimate. It is interesting to observe that the location estimates from the unsynchronized and unsynchronized negative-sequence two-ended methods show a significant improvement over the location estimates from the one-ended methods (see previous example). Recall that an inaccurate value of the zero-sequence line impedance was considered the most likely suspect for the fault location error in the one-ended methods. Because two-ended methods do not use zero-sequence line impedance in their fault location calculation, estimates are accurate. In summary, this example highlights the superior performance of two-ended methods over one-ended methods in computing the distance to a fault.

■ Exercise 3.4

A lightning strike caused a three-phase fault on a two-terminal transmission line with a nominal voltage rating of 161 kV. The transmission line between terminal G and terminal H is 83.45 miles long and has a positive-sequence impedance of $Z_1 = 9.4612 + j63.6878\ \Omega$. Linemen located the fault at a distance of 36.36 miles from terminal G as shown in Figure 3.23. Voltage and current waveforms captured by a digital fault recorder (DFR) at terminal G during the event are shown in Figure 3.24. Sampling rate of the DFR is 100 samples per cycle. The prefault current and voltage phasors at cycle 15.4 and the fault current and voltage phasors at cycle 19.4 are given below.

$$
\begin{array}{cccc}
I_{G_{abc}}\ [\text{kA}] & V_{G_{abc}}\ [\text{kV}] & I^f_{G_{abc}}\ [\text{kA}] & V^f_{G_{abc}}\ [\text{kV}] \\
\begin{bmatrix} 0.19\angle-130.30° \\ 0.21\angle108.91° \\ 0.20\angle-14.36° \end{bmatrix} &
\begin{bmatrix} 95.25\angle-140.66° \\ 94.87\angle98.99° \\ 94.20\angle-20.84° \end{bmatrix} &
\begin{bmatrix} 3.10\angle142.40° \\ 3.18\angle18.21° \\ 3.05\angle-104.25° \end{bmatrix} &
\begin{bmatrix} 85.05\angle-140.60° \\ 84.51\angle99.63° \\ 83.76\angle-20.16° \end{bmatrix}.
\end{array}
$$

A DFR at terminal H also recorded the voltage and current waveforms during the event at 96 samples per cycle that are shown in Figure 3.25. The prefault current and voltage phasors at cycle 8.4 and the fault current and voltage phasors at cycle 11.4 are given below.

$$
\begin{array}{cccc}
I_{H_{abc}}\ [\text{kA}] & V_{H_{abc}}\ [\text{kV}] & I^f_{H_{abc}}\ [\text{kA}] & V^f_{H_{abc}}\ [\text{kV}] \\
\begin{bmatrix} 0.18\angle41.96° \\ 0.19\angle-79.39° \\ 0.17\angle157.41° \end{bmatrix} &
\begin{bmatrix} 97.25\angle-143.44° \\ 96.61\angle96.78° \\ 96.37\angle-24.56° \end{bmatrix} &
\begin{bmatrix} 2.28\angle138.93° \\ 2.35\angle14.57° \\ 2.19\angle-106.99° \end{bmatrix} &
\begin{bmatrix} 86.79\angle-144.15° \\ 85.68\angle96.39° \\ 85.87\angle-24.67° \end{bmatrix}.
\end{array}
$$

The fault was cleared quickly within three cycles at both line terminals. Using the information captured by the DFRs, determine the following:

(a) Source impedance of terminal G.
(b) Source impedance of terminal H.
(c) Fault location using the simple reactance method.
(d) Fault location using the Takagi method.
(e) Fault location using the current distribution factor method.

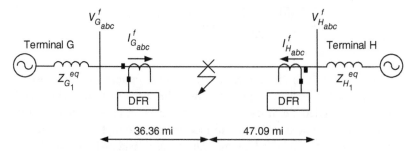

Figure 3.23 Exercise 3.4: One-line showing the location of the three-phase fault.

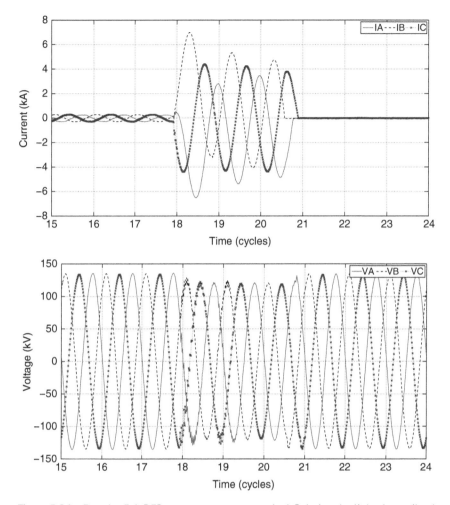

Figure 3.24 Exercise 3.4: DFR measurements at terminal G during the lightning strike that resulted in a three-phase fault.

Solution:

(a) Because this is a three-phase fault, we can estimate the positive-sequence source impedance behind terminal G using (3.24):

$$Z_{G_1}^{eq} = -\frac{V_{G_1}^f - V_{G_1}}{I_{G_1}^f - I_{G_1}} \quad [\Omega].$$

Now, the positive-sequence voltage and current in a three-phase fault is nothing but the voltage and current measured in any one of the three phases. We choose phase C since this phase has the least amount of DC offset as seen in Figure 3.24. Rewriting the above equation in terms of C-phase quantities, we can calculate the source impedance

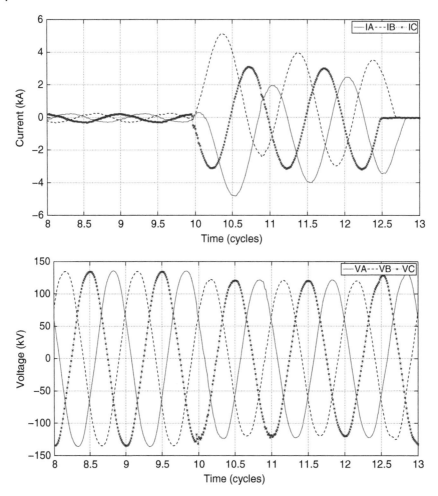

Figure 3.25 Exercise 3.4: DFR measurements at terminal H during the lightning strike that resulted in a three-phase fault.

of terminal G:

$$Z_{G_1}^{eq} = -\frac{V_{G_c}^f - V_{G_c}}{I_{G_c}^f - I_{G_c}}$$

$$= -\frac{83.76\angle - 20.16° - 94.20\angle - 20.84°}{3.05\angle - 104.25° - 0.20\angle - 14.36°}$$

$$= 0.4941 + j3.3977 \quad [\Omega].$$

(b) Waveform data captured by the DFR at terminal H, shown in Figure 3.25, indicate that the DC offset in phase C is almost zero. Therefore, substituting the C-phase quantities in place of the positive-sequence quantities in (3.24), we can calculate the source impedance of terminal H:

$$Z^{eq}_{H_1} = -\frac{V^f_{H_c} - V_{H_c}}{I^f_{H_c} - I_{H_c}}$$

$$= -\frac{85.87\angle - 24.67° - 96.37\angle - 24.56°}{2.19\angle - 106.99° - 0.17\angle 157.41°}$$

$$= 0.9099 + j4.6572 \quad [\Omega].$$

It is interesting to note that the system is fairly homogeneous as the local terminal, the remote terminal, and the transmission line have similar impedance angles of 81.72°, 78.94°, and 81.55°, respectively.

(c) We apply the simple reactance method on data captured by the DFR at terminal G to obtain a location estimate. First, we need to determine V_G and I_G from Table 3.1. For a three-phase fault, V_G (and I_G) is the difference in voltage (and current) between two phases. In this example, we choose phases C and A since the DC offset in these phases is lower than that in phase B. V_G is calculated as below.

$$V_G = V^f_{G_c} - V^f_{G_a}$$

$$= 83.76\angle - 20.16° - 85.05\angle - 140.60°$$

$$= 146.52\angle 9.87° \quad [kV].$$

I_G is calculated as below.

$$I_G = I^f_{G_c} - I^f_{G_a}$$

$$= 3.05\angle - 104.25° - 3.10\angle 142.40°$$

$$= 5.14\angle - 70.57° \quad [kA].$$

Next, we substitute V_G and I_G in (3.10) and calculate the distance to fault:

$$m = \frac{\text{imag}\left(\dfrac{V_G}{I_G}\right)}{\text{imag}\left(Z_1\right)}$$

$$= \frac{\text{imag}\left(\dfrac{146.52\angle 9.87°}{5.14\angle - 70.57°}\right)}{63.6878}$$

$$= 0.44 \quad [pu].$$

To obtain the fault location in units of miles, we multiply m with the line length and get 36.72 miles. The actual fault location is 36.36 miles from terminal G.

(d) The Takagi method uses the pure fault current to minimize any fault location error due to the combined effect of load and fault resistance. Table 3.2, the pure fault current for a three-phase fault can be calculated from terminal G data:

$$\Delta I_G = \left(I^f_{G_c} - I_{G_c}\right) - \left(I^f_{G_a} - I_{G_a}\right)$$

$$= \left(3.05\angle - 104.25° - 0.20\angle - 14.36°\right) - \left(3.10\angle 142.40° - 0.19\angle - 130.30°\right)$$

$$= 5.13\angle - 74.31° \quad [kA].$$

Next, we calculate the fault location:

$$m = \frac{\text{imag}\left(V_G \times \Delta I_G^*\right)}{\text{imag}\left(Z_1 \times I_G \times \Delta I_G^*\right)}$$

$$= \frac{\text{imag}\left(146.52\angle 9.87° \times 5.13\angle 74.31°\right)}{\text{imag}\left((9.4612 + j63.6878) \times 5.14\angle - 70.57° \times 5.13\angle 74.31°\right)}$$

$$= 0.44 \quad [\text{pu}].$$

Multiplying with the line length yields a location estimate of 36.72 miles from terminal G.

(e) To implement the current distribution factor method, our first task is to calculate constants a, b, c, d, e, and f defined in Section 3.1.4. Constants a and b are calculated as
follows:

$$a + jb = 1 + \frac{Z_{H_1}^{eq}}{Z_1} + \left(\frac{V_G}{Z_1 \times I_G}\right)$$

$$= 1 + \frac{0.9099 + j4.6572}{9.4612 + j63.6878} + \left(\frac{146.52\angle 9.87°}{(9.4612 + j63.6878) \times 5.14\angle - 70.57°}\right)$$

$$= 1.5160 - j0.0118.$$

Constants c and d are calculated as follows:

$$c + jd = \frac{V_G}{Z_1 \times I_G}\left(1 + \frac{Z_{H_1}^{eq}}{Z_1}\right)$$

$$= \frac{146.52\angle 9.87°}{(9.4612 + j63.6878) \times 5.14\angle - 70.57°}\left(1 + \frac{0.9099 + j4.6572}{9.4612 + j63.6878}\right)$$

$$= 0.4750 - j0.0106.$$

Constants e and f are calculated as follows:

$$e + jf = \frac{\Delta I_G}{Z_1 \times I_G}\left(1 + \frac{Z_{H_1}^{eq} + Z_{G_1}^{eq}}{Z_1}\right)$$

$$= \frac{5.13\angle - 74.31°}{(9.4612 + j63.6878) \times 5.14\angle - 70.57°}\left(1 + \frac{1.4040 + j8.0549}{9.4612 + j63.6878}\right)$$

$$= 0.0014 - j0.0174.$$

After substituting the constants in (3.22), we solve the quadratic equation and obtain
two values of m.

$$m = \frac{\left(a - \frac{eb}{f}\right) \pm \sqrt{\left(a - \frac{eb}{f}\right)^2 - 4\left(c - \frac{ed}{f}\right)}}{2}$$

$$= \frac{1.5151 \pm \sqrt{0.3991}}{2}$$

$$= 0.44, 1.07 \quad [\text{pu}].$$

Since m needs to have a value between 0 and 1 per unit, 0.44 per unit (36.72 miles)
is chosen as the location estimate. It is interesting to observe that the Takagi and the

current distribution factor method did not offer any significant improvement over the simple reactance method in this example. In fact, all the methods gave identical estimates. This is surprising as the load current was significant, around 200 A. The absence of any reactance error due to load suggests that the fault occurred with negligible fault resistance.

■ Exercise 3.5

A B-G fault occurred on a 69 kV transmission line between Bus G and Bus H. The line is 72.77 miles long. A traveling-wave fault locator at Bus G saw the first wave at 8:39:36.832684476 p.m. while a traveling-wave fault locator at Bus H saw the first wave at 8:39:36.832667109 p.m. The utility had done a line energization test when commissioning the fault locators and had determined the line propagation velocity of the waves to be 0.98821 times the speed of light. Unfortunately, because the communication channel between the fault locators was offline at the time of the fault, the fault locators were not able to calculate a fault location automatically. Use the given information to manually calculate a fault location using the double-ended traveling wave method.

Solution:

The traveling-wave fault locator at Bus H saw the wave before the traveling-wave fault locator at Bus G. This means that the fault was closer to Bus H. Using (3.71), the fault location can be calculated as:

$$
m = \frac{v_p \times c \times (T_{G1} - T_{H1}) + LL}{2}
$$

$$
= \frac{0.98821 \times 186,282.397 \times 17.37 \times 10^{-6} + 72.77}{2}
$$

$$
= 37.98 \ [\text{mi}].
$$

The actual fault is at 37.97 miles from Bus G. A broken insulator was responsible for the fault. Notice how easily we were able to calculate the fault location. Plus the accuracy was spot on!

3.6   Summary

This chapter presents the theory behind one- and two-ended impedance-based fault location algorithms. The data required by each algorithm is summarized below. The next chapter will discuss the error sources that affect the accuracy of each method. We also discussed impedance-based algorithms that can locate faults on three-terminal lines. Finally, we discussed traveling-wave algorithms.

Input Data	Simple Reactance	Takagi	Modified Takagi	Current Distribution Factor	Synchro-nized	Unsyn-chronized	Unsynchronized Negative-sequence	Synchronized Line Current Differential
Fault Event Data								
Fault type	✓	✓	✓	✓				✓
Fault voltage (local end)	✓	✓	✓	✓	✓	✓		✓
Fault current (local end)	✓	✓	✓	✓	✓	✓	✓	✓
Fault voltage (remote ends)					✓	✓		
Fault current (remote ends)					✓	✓	✓	✓
Synchronized data					✓			✓
Prefault current		✓		✓				
Line Parameters								
Line length	✓	✓	✓	✓	✓	✓	✓	✓
Positive-sequence Line impedance	✓	✓	✓	✓	✓	✓	✓	✓
Zero-sequence Line impedance	✓	✓	✓	✓				✓
Source Impedance Parameters								
Positive-sequence Source impedance (local end)			✓	✓			✓	
Positive-sequence Source impedance (remote end)			✓	✓			✓	
Zero-sequence Source impedance (local end)			✓					
Zero-sequence Source impedance (remote end)			✓					

4

Error Sources in Impedance-Based Fault Location

Impedance-based fault location algorithms described in Chapter 3 require the inputs of voltage and current phasors during a fault, line impedance and line length, and fault type when estimating the fault location. Inaccurate input data can result in an inaccurate location estimate. In addition, impedance-based fault location algorithms make certain simplifying assumptions when computing the distance to a fault. Assumptions made include no load at the time of the fault, no fault resistance, no parallel lines, and homogeneous system to name a few. Accuracy is affected when the algorithms are applied to a system in which these assumptions do not hold true. Because of the incorrect application, the assumptions become sources of error. In this chapter, we use a power system model to show the sensitivity of fault-locating algorithms to errors stemming from inaccurate input data and incorrect application. Understanding the impact of these error sources on fault location will help the user make decisions in the field when a fault occurs in their system.

4.1 Power System Model

We modeled a two-terminal transmission network in an electromagnetic transients power system simulation software package. We used this model to replicate actual faults on a transmission line and generate the corresponding voltage and current waveforms that would be captured at the line terminals. We set up the model to capture waveforms at a high resolution of 128 samples per cycle. The system nominal voltage is 69 kV. Refer to Figure 3.1 for the one-line representation of the system. The network upstream from terminal G was represented by an ideal voltage source $E_G = 1\angle 10°$ per unit and an equivalent positive- and zero-sequence impedance of $Z_{G_1}^{eq} = 3.75\angle 71°\ \Omega$ and $Z_{G_0}^{eq} = 11.25\angle 65°\ \Omega$, respectively. The network upstream from terminal H was also represented by an ideal voltage source $E_H = 1\angle 0°$ per unit and an equivalent positive- and zero-sequence impedance of $Z_{H_1}^{eq} = 12\angle 71°\ \Omega$ and $Z_{H_0}^{eq} = 30\angle 65°\ \Omega$, respectively. The angle by which E_G leads E_H is known as the power angle, δ, and represents the net load served by the transmission line. The transmission line connecting the two terminals is 18 miles long and was modeled with the frequency dependent model, which is the most accurate method by which to model a transmission line. We used the tower configuration of an actual 69 kV transmission line as shown in Figure 4.1. Shield wires S_1 and S_2 protect phase conductors A, B, and C

Fault Location on Transmission and Distribution Lines: Principles and Applications, First Edition.
Swagata Das, Surya Santoso, and Sundaravaradan N. Ananthan.
© 2022 John Wiley & Sons Ltd. Published 2022 by John Wiley & Sons Ltd.
Companion website: www.wiley.com/go/das/faultlocation

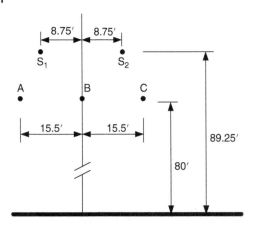

Figure 4.1 69 kV tower configuration.

Table 4.1 Conductor Data

	Material	Resistance (Ω/mi)	Diameter (inch)	GMR (feet)
Phase	ACSR Linnet 336,400 26/7	0.294	0.720	0.024
Shield	ACSR Grouse 80,000 8/1	1.404	0.367	0.009

from direct lightning strikes. The conductor material is described in Table 4.1. Note that we intentionally designed the model to be simple, homogeneous, and compliant with all the assumptions made by impedance-based fault location algorithms. The goal was to introduce the fault-locating error sources one by one and study the impact on fault location estimates. Since a simple model was being used, the error in location estimates would be strictly proportional to the inaccuracies introduced and would, therefore, give an accurate measure of how significant a particular error source is and whether the error source should be considered for fault location purposes.

4.2 Input Data Errors

Impedance-based fault location algorithms require the inputs of voltage and current phasors during a fault to calculate the fault location. Unfortunately, because of DC offset and CT saturation, fault current phasors may not be accurate. Because of aging coupling-capacitor voltage transformers (CCVTs) and open-delta connected voltage transformers, fault voltage phasors may not be accurate. In addition, all impedance-based fault location algorithms estimate the impedance to fault in ohms. Line impedance parameters in ohms per unit distance are required to obtain a corresponding distance estimate. Uncertainty about these line parameters, particularly the zero-sequence line impedance, further adds to the error in location estimates. The equation used by some fault location algorithms depends on fault type. If the fault type is chosen incorrectly, fault location will be inaccurate. This section discusses these error sources and their impact on fault location accuracy in further detail. You will see that the accuracy of the fault location algorithms is as good as the input data.

4.2.1 DC Offset

Impedance-based fault location algorithms calculate the impedance to fault by using voltage and current signals measured during the fault. Dividing the result with the line impedance per unit line length converts the impedance estimate to a distance estimate. Now this impedance is calculated at a certain frequency, typically 60 Hz in North America. This means that the only frequency that is of interest in the voltage and current signals is 60 Hz. All other frequencies are noise and corrupt the data, which means that they must be removed from the measured voltage and current signals before being used to estimate the impedance to a fault.

DC offset is one such noise that must be removed from the measured current waveforms. You can think of the power system as a resistive-inductive system. When a fault occurs in this resistive-inductive system, the AC symmetrical fault current is often accompanied with an exponentially decaying DC offset. The DC offset makes the fault current asymmetrical as shown in Figure 4.2 during an A-G fault. It shows up to satisfy the two conflicting laws that

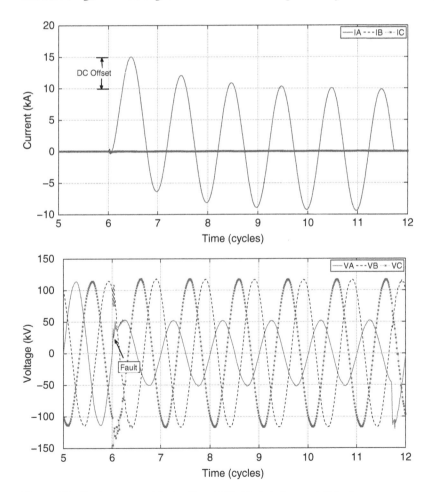

Figure 4.2 Fault current with significant DC offset during an A-G fault.

govern our power system. First, current must lag voltage by the natural power factor of the system. So if a fault occurs at the zero-crossing of a voltage waveform in a purely inductive system, current must lag the voltage by 90 degrees, meaning it must start at the maximum negative peak. Second, current cannot change instantaneously in an inductive system.

Let us break down the above explanation with the waveforms shown in Figure 4.2. Notice that the A-G fault occurred close to the zero-crossing of the A-phase voltage. At that point, the A-phase current was expected to be close to the maximum negative-peak according to the first law. However, because current cannot go from zero to its maximum value instantaneously in an inductive system according to the second law, a counter current, which is DC in nature, was produced that resulted in the total current (AC symmetrical current plus dc current) starting from zero. This DC current was equal and opposite to the AC current required to satisfy the first law and decayed with a time constant equal to the X/R ratio of the system. The DC current is maximum when a fault occurs at the zero-crossing of a voltage waveform and minimum when a fault occurs near the voltage peak. Most single line-to-ground faults are caused by insulation breakdown due to animal or tree contact during peak voltage conditions and result in a fault current with negligible DC offset [71]. Faults due to lightning strikes are, however, random and can occur at any point on the voltage waveform.

Fault locators must use a digital filter to reject the DC offset and extract the 60 Hz component of the current signal. The fast Fourier filter (FFT filter) is a popular filter [72]. It uses a window length of one cycle to extract the 60 Hz magnitude and phase angle of a signal and discard the DC offset and harmonics. To illustrate the performance of this filter, we applied the one-cycle Fourier filter to the current waveform in Figure 4.2. We plotted the 60 Hz current magnitude, which is one of the outputs of the filter, in Figure 4.3. You can see that the fast Fourier filter was successful in filtering out most, but not all, of the DC offset. The fault current magnitude oscillated about the actual current magnitude and reached steady state only after the DC offset decayed out. Because the fault current magnitude fluctuated, the location estimate from the simple reactance method also fluctuated as shown in Figure 4.4. Whenever the current magnitude output from the filter was greater than the actual current

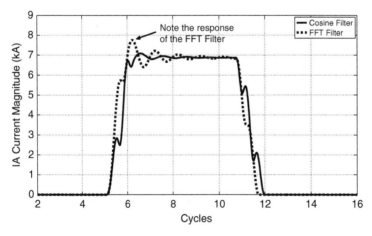

Figure 4.3 Cosine filter more effective in filtering out the DC offset.

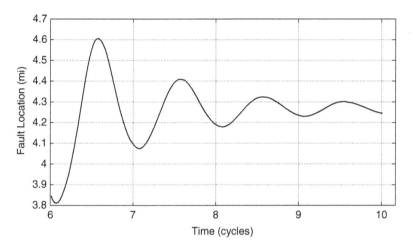

Figure 4.4 Impact of DC offset on fault location estimates from the simple reactance method. Voltage and current phasors were extracted using an FFT filter.

magnitude, the simple reactance method overestimated the fault location. Whenever the current magnitude output from the filter was lower than the actual current magnitude, the simple reactance method underestimated the fault location.

The cosine filter is another commonly used filter. The coefficients of this filter are sampled from a cosine wave and require a minimum response time of one and one-quarter cycles. The quarter-cycle delay is used to calculate the phase angle. As seen in Figure 4.3, the cosine filter does a better job of eliminating the DC offset than the FFT filter [73]. With both filters, however, the front and tail end of the signal are severely distorted. When the fault first occurs, it takes about a cycle for the Fourier and one and one-quarter cycle for the cosine filter to build up to the fault current magnitude. Similarly, when the breaker opens to interrupt the fault, it takes the same amount of time to flush out the fault data and go to zero. Using current data from the front and tail end of the fault can, therefore, result in erroneous location estimates. It is best to use data from the middle of the fault to calculate fault location.

4.2.2 CT Saturation

Fault locators rely on current transformers to step down the primary current and to produce a secondary current that closely replicates the primary current. Figure 4.5 shows a CT measuring current in a 138 kV substation. The CT secondary current output goes through secondary cabling to relays, fault locators, and DFRs in the control house. Unfortunately, CTs can saturate during a fault and produce a signal that is not representative of the primary current. The erroneous current if used in fault location calculations will certainly result in a significant fault location error.

CTs can saturate if they haven't been sized correctly [74]. The high magnitude fault current may be too large for the CT iron core to handle, causing it to saturate. This is known as symmetrical saturation and is shown in Figure 4.6. The continuous line in black, marked as ideal, shows the primary current. The dashed line in black shows the output of a saturated CT. Note that the secondary signal was scaled by the CT ratio before being plotted against

Figure 4.5 CT measuring current in a 138 kV substation.

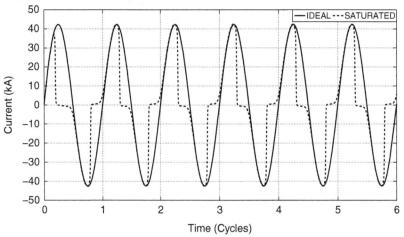

Figure 4.6 Symmetrical CT saturation.

the primary current to make the comparison easier. The CT starts off by exactly replicating the primary current. However, before the primary current can reach its peak, the CT core saturates. At this point, even though there is current flow in the primary, the output of the CT becomes zero until the primary current starts flowing in the negative direction. The CT again starts to exactly replicate the primary current until the core saturates and the output of the CT drops to zero. This results in a characteristic sawtooth waveform. Reference [75] provides guidance on how to size a CT correctly. Avoid performing fault location calculations if you suspect symmetrical saturation on inspecting the current waveforms.

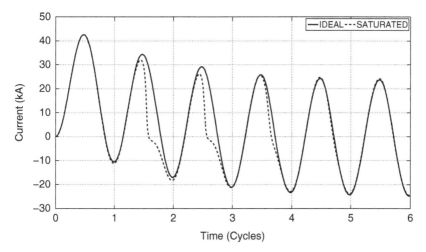

Figure 4.7 Asymmetrical CT saturation.

CTs can also saturate if there is a significant DC offset. The current offset in either the positive or the negative direction causes the CT to saturate and is referred to as asymmetrical saturation. An example of this is shown in Figure 4.7. Notice that after the DC offset decays down within two or three cycles, the CT comes out of saturation. Therefore, for faults that last for a number of cycles, the best way to handle asymmetrical CT saturation is to wait for the DC offset to decay before applying the fault location algorithms.

4.2.3 Aging CCVTs

Voltage transformers (VTs), also referred to as potential transformers (PTs), measure the primary system voltage, step it down to a safe secondary voltage level, and bring that information to fault locators, relays, or DFRs. They can be of two types: (1) wire-wound VTs or (2) coupling-capacitor VTs (also known as CCVTs). CCVTs are more economical than wire-wound VTs and are typically used to measure transmission line voltages, 138 kV or above [76]. A CCVT measuring voltage in a 138 kV substation is shown in Figure 4.8. Inside the housing, capacitors are used to create a voltage divider and step down the voltage. A tuning reactor is used to cancel the phase shift between the primary and the secondary voltages [77]. The secondary voltage is then stepped down even further to 66.4 V secondary by a wire-wound VT.

As you can tell from the above discussion, the construction of a CCVT is complex and has many components. With age, the probability of one of those components to fail increases. The failure can manifest itself as a total loss of secondary voltage, a secondary signal that does not replicate the primary voltage, or an unexpected transient. Since voltage is an input to all but the unsynchronized negative-sequence two-ended method, an erroneous voltage signal will cause a significant error in fault location.

4.2.4 Open-Delta VTs

Utilities have two choices when connecting VTs to a fault locator. The first choice is to use three single-phase PTs and connect them in a wye configuration as shown in Figure 4.9(a).

Figure 4.8 CCVTs measuring line-to-ground bus voltages in a 138 kV substation.

This connection type is the most common and provides the fault locator with line-to-ground voltage measurements. The second choice is to use two single-phase PTs and connect them in an open-delta configuration as shown in Figure 4.9(b). In this configuration, the fault locator measures zero volts at VB input (this is grounded), V_{ab} at the VA input, and V_{cb} at the VC input. The fault locator can calculate V_{ca} as:

$$V_{ab} + V_{bc} + V_{ca} = 0$$
$$\therefore V_{ca} = -\left(V_{ab} + V_{bc}\right) = V_{cb} - V_{ab}.$$

(4.1)

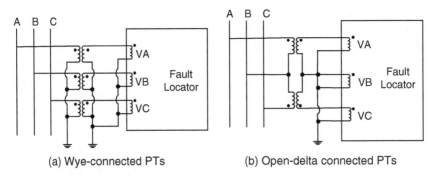

(a) Wye-connected PTs   (b) Open-delta connected PTs

Figure 4.9 Types of VT connection.

The open-delta connection is more economical as it saves utilities the cost of a third PT. The measured line-to-line voltages can be directly used by single-ended methods to estimate the location of line-to-line, double line-to-ground, and three-phase faults as shown in Table 3.1. However, the savings come at the cost of the zero-sequence voltage, which is filtered out. The unavailability of the zero-sequence voltage poses a challenge to single-ended methods that require the line-to-ground voltage of the faulted phase when locating single line-to-ground faults. To calculate line-to-ground voltage, you need to sum positive-, negative-, and zero-sequence voltages. While the positive- and negative-sequence voltages can be easily calculated from line-to-line voltages, the challenge lies in the fact that the zero-sequence voltage is not available.

Reference [54] suggests estimating the faulted phase-to-ground voltage if the zero-sequence source impedance is known. For example, suppose there is an A-G fault on the transmission line shown in Figure 3.1. The fault locator at terminal G is measuring line-to-line voltage measurements through an open-delta connected PT. The fault locator requires the faulted phase voltage, $V^f_{G_a}$, to calculate the location of the A-G fault. We can express this voltage in terms of symmetrical components as:

$$V^f_{G_a} = V^f_{G_1} + V^f_{G_2} + V^f_{G_0}. \tag{4.2}$$

The positive-sequence voltage, $V^f_{G_1}$, can be calculated from line-to-line voltages for a system with ABC phase rotation as:

$$V^f_{G_1} = \frac{1}{3\sqrt{3}} \left[V^f_{G_{ab}} + aV^f_{G_{bc}} + a^2V^f_{G_{ca}} \right] \times e^{-j30°}. \tag{4.3}$$

The negative-sequence voltage, $V^f_{G_2}$, can be calculated from line-to-line voltages for a system with ABC phase rotation as:

$$V^f_{G_2} = \frac{1}{3\sqrt{3}} \left[V^f_{G_{ab}} + a^2V^f_{G_{bc}} + aV^f_{G_{ca}} \right] \times e^{j30°}. \tag{4.4}$$

Because the zero-sequence voltage is not available with the open-delta PT configuration, it must be estimated. In the zero-sequence network shown in Figure 3.2, the zero-sequence voltage during a ground fault is nothing but the voltage drop across the zero-sequence source impedance and can be written as:

$$V^f_{G_0} = -Z^{eq}_{G_0} \times I^f_{G_0}. \tag{4.5}$$

Summing (4.3), (4.4), and (4.5), $V^f_{G_a}$ can be estimated as

$$V^f_{G_{a-est}} = \frac{1}{3} \left(V^f_{G_{ab}} - V^f_{G_{ca}} \right) - Z^{eq}_{G_0} I^f_{G_0}. \tag{4.6}$$

In a similar manner, the faulted phase voltage during a B-G fault can be estimated as:

$$V^f_{G_{b-est}} = \frac{1}{3} \left(V^f_{G_{bc}} - V^f_{G_{ab}} \right) - Z^{eq}_{G_0} I^f_{G_0}. \tag{4.7}$$

The faulted phase voltage during a C-G fault can be estimated as:

$$V^f_{G_{c-est}} = \frac{1}{3} \left(V^f_{G_{ca}} - V^f_{G_{bc}} \right) - Z^{eq}_{G_0} I^f_{G_0}. \tag{4.8}$$

The above equations are true for a system with ABC or ACB phase rotation. Accuracy of the estimated line-to-ground fault voltage depends on the accuracy of the zero-sequence source impedance.

4.2.5 Inaccurate Line Length

All impedance-based fault location algorithms require the input of line length to convert the apparent impedance to fault to a distance estimate. The definition of line length can be quite ambiguous and could be any one of the following: (1) the actual geographical distance between two substations, (2) the straight line distance between two substations neglecting terrain (obtained from a map), (3) the straight line distance between two substations including terrain, or (4) length of the transmission line conductor including sag [63]. The last method is the most accurate way of defining line length but can be difficult to measure accurately since it changes with ambient temperature and load. This ambiguity about line length and difficulty to measure it accurately can also result in a fault location error.

4.2.6 Untransposed Lines

Impedance-based fault location algorithms require the positive- and zero-sequence impedances of a transmission line to estimate the distance to a fault. When calculating the sequence line parameters, transmission lines are assumed to be transposed. Transposition is the principle of physically exchanging the position of phase conductors at periodic intervals such that a particular conductor occupies all positions of a particular line configuration. This principle is illustrated in Figure 4.10, where the positions of phase conductors A, B, and C are rotated every one-third of the total line length.

Transposition equalizes the mutual coupling between the three phases and reduces the sequence impedance matrix, Z_{012}, to a diagonal matrix as shown in (2.33). The diagonal elements are formed by the sequence line impedances while the off-diagonal elements are zero, indicating no coupling between the sequence networks.

Although transposing a line is advantageous, it introduces complications in the design of a transmission line and increases the overall design cost due to additional support structures and insulator string requirements. The probability of faults also increase when the conductors cross over. As a result, many transmission lines are not transposed.

We wanted to investigate the impact of untransposed lines on the accuracy of impedance-based fault location algorithms. To do this, we took our power system model and simulated single line-to-ground faults along the length of the transmission line with $R_f = 0\,\Omega$. This serves as the reference case as the transmission line is transposed. We calculated distances to faults by applying single-ended methods to the voltage and current measurements recorded at terminal G and two-ended methods to the measurements captured at both line terminals. Next, we took the same transmission line in our power system test model and

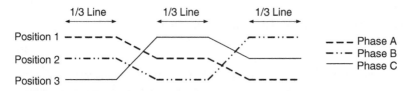

Figure 4.10 A transposed transmission line [78].

made it untransposed. The sequence impedance matrix is shown below.

$$Z_{012} = \begin{bmatrix} 15.47 + j32.52 & 0.26 + j0.00 & 0.26 + j0.00 \\ 0.26 + j0.00 & 5.33 + j15.15 & 0.00 - j1.02 \\ 0.26 + j0.00 & 0.00 - j1.02 & 5.33 + j15.15 \end{bmatrix} \Omega.$$

You can see that the off-diagonal elements are no longer zero and represent the coupling between the sequence networks. We simulated the same faults with the same value of R_f, and calculated the distances to faults using the new set of voltage and current waveforms. The results are shown in Figure 4.11. The fault location error was calculated using (1.1).

Because impedance-based fault location algorithms assume that the line is transposed and that the sequence networks are decoupled from each other, an untransposed line affected the accuracy of the location estimates. Single-ended methods underestimated the location of a fault when compared against the reference case. The fault-location error increases as faults move farther away from the monitoring location. Two-ended methods

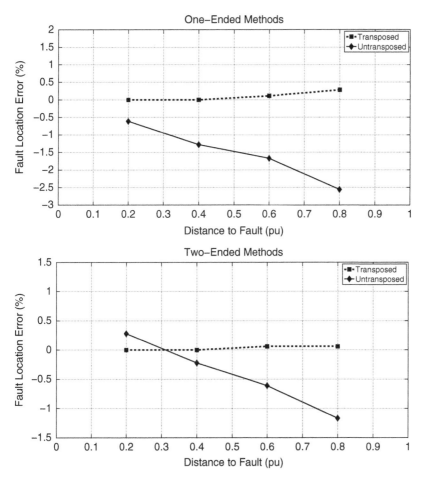

Figure 4.11 Fault location error due to untransposed transmission lines.

are also affected by the line transposition assumption, the fault location error being around 1.2%.

4.2.7 Variation in Earth Resistivity

Earth resistivity (ρ) is the resistance with which the earth opposes the flow of ground current. It is an electrical characteristic of the ground and is used when calculating the zero-sequence impedance of a transmission line [45]. Determining the exact value of ρ is difficult since it varies greatly with the soil type as shown in Table 4.2. Most utilities use a standard earth resistivity value of 100 Ω-m while others use the Wenner four-point method to measure ρ for greater accuracy [55]. In addition to soil type, the value of ρ is also dictated by the moisture content in soils, temperature, and season of the year. Under extremely high or low temperatures, the soil is dry and has a very high resistivity. During the rainy season, the value of ρ decreases. Minerals, salts, and other electrolytes make soils more conductive and tend to lower the resistivity. As a result, earth resistivity is not constant and is not known accurately.

Table 4.3 shows the impact of a varying earth resistivity value on the positive- and zero-sequence impedances of the transmission line in the power system test model. As expected, the positive-sequence line impedance remains unaffected by changes in the value of earth resistivity. The zero-sequence line impedance, on the other hand, increases as earth resistivity increases. All single-ended fault location algorithms require the zero-sequence line impedance to compute the location of single line-to-ground faults and are sensitive to any changes in earth resistivity. One of the two-ended methods, the synchronized line current differential method, also makes use of the zero-sequence line impedance and is affected by the variation in earth resistivity.

We use the power system test model to study the impact of an unknown value of earth resistivity on impedance-based fault location algorithms. We simulated single line-to-ground faults along the entire length of the transmission line. At each fault location, we varied the earth resistivity from 10 to 1000 Ω-m. The zero-sequence line impedance, used as an input to the fault location algorithms, was, however, calculated

Table 4.2 Variation of Earth Resistivity with Soil Type [79]

Soil Type	Earth Resistivity (Ω-m)	
	Range	Average
Peat	>1200	200
Adobe clay	2–200	40
Boggy ground	2–50	30
Gravel (moist)	50–3000	1000 (moist)
Sand and sandy ground	50–3000	200 (moist)
Stony and rocky ground	100–8000	2000
Concrete: 1 part cement + 3 parts sand	50–300	150

Table 4.3 Effect of Earth Resistivity on Line Impedance
Parameters

ρ (Ω-m)	Z_1 (Ω)	Z_0 (Ω)
10	$5.33 + j15.15$	$13.59 + j30.34$
100	$5.33 + j15.15$	$15.47 + j32.53$
500	$5.33 + j15.15$	$16.74 + j33.87$
1000	$5.33 + j15.15$	$17.28 + j34.40$

(ρ increases ↓) (Z_{L0} increases ↓)

using the standard earth resistivity value of 100 Ω-m. The test replicates a real-world scenario in which the actual earth resistivity value can be different than the one used to calculate the line impedance settings of the fault locator.

The results are shown in Figure 4.12. The reference case is the one in which the actual earth resistivity matched the one used to calculate the zero-sequence line impedance. You can see that the accuracy of the single-ended methods was affected by the uncertainty in earth resistivity. When the actual value of earth resistivity was greater than the one used in the fault location computation, i.e., 100 Ω-m, the distance to fault was overestimated. Similarly, when the earth resistivity was lower than the value used in the fault location computation, the distance to fault was underestimated. The fault location error increased when faults were farther away from the monitoring location. This is because the error due to inaccurate zero-sequence line impedance added up as the distance to fault increased. Among the two-ended methods, the synchronized line current differential will be affected by the uncertainty in earth resistivity values as explained earlier. In contrast, the other two-ended methods described in Chapter 3 do not use zero-sequence line impedance when estimating the distance to fault and hence were not affected by the variation in earth resistivity as shown in Figure 4.12.

4.2.8   Non-Homogeneous Lines

Both one- and two-ended impedance-based fault location algorithms assume that the transmission line is homogeneous with a uniform impedance per mile. This assumption may not always be true and will result in an error in fault location [40]. For example, conductor size or material may change as the line goes from one substation to the other. Line construction, i.e., the arrangement of the phase conductors on the pole, may also change along the way. For example, Figure 4.19 shows a horizontal configuration and Figure 4.20 shows a vertical arrangement of the phase conductors. Tower structures at the entrance or exit to a substation may be different than line structures. Earth resistivity may change along the line length, resulting in the line having a non-homogeneous zero-sequence impedance. Section 5.2.2 discusses different approaches to reduce fault location error stemming from applying impedance-based fault location algorithms to a non-homogeneous line.

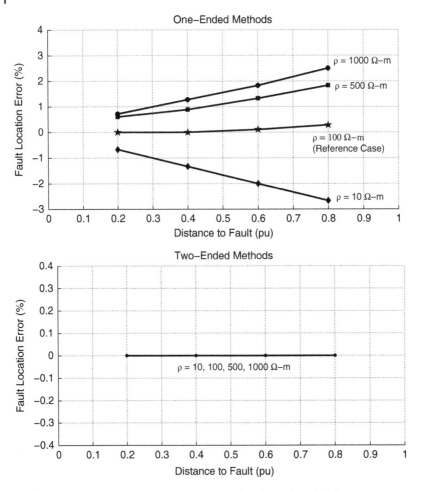

Figure 4.12 Fault location error due to uncertainty in earth resistivity.

Transmission lines can also be non-homogeneous if they consist of overhead and underground line sections, which are known as hybrid lines [63, 80]. Underground line sections are quite expensive to construct but help in crossing through airports, railway tracks, and other highly populated areas where obtaining a right-of-way to build overhead lines may be difficult. Impedance-based fault location algorithms will not be successful in locating faults on hybrid lines. First, the degree of non-homogeneity is quite significant as the positive-sequence impedance per mile of the overhead section is very different than the positive-sequence impedance per mile of the underground section. Second, the zero-sequence line impedance of the underground section is a nonlinear function of the fault current [81]. Finally, the zero-sequence line impedance of the underground section is uncertain as it depends on the grounding of the cable shield and presence of other parallel conductive paths such as gas and water pipes. As a result, all single-ended methods that use the zero-sequence line impedance to locate ground faults will be inaccurate. All the two-ended methods except for the synchronized line current differential method will not be affected by the error in zero-sequence line impedance but will be affected by the significant non-homogeneity of the hybrid line.

4.2.9   Incorrect Fault Type Selection

All of the single-ended fault location algorithms and the synchronized line current differential algorithm (a two-ended algorithm) require knowledge of the fault type as described in Chapter 3. Automatic fault locators that make use of these algorithms have a fault-type selection logic to first determine the fault type. After the fault type is identified, the distance to fault is calculated using the phase or phases involved in the fault. The fault-type selection logic maybe something as simple as a phase overcurrent check. Another method runs all six distance mho calculations (AG, BG, CG, AB, BC, and CA). The loop whose calculation plots inside the mho impedance circle and is closer to the origin is selected as the fault type [82]. Other methods compare the relationship between the zero- and negative-sequence current or between the positive- and negative-sequence current to determine fault type as explained in [83]. Unfortunately, a weak source can challenge the fault selection logic inside automatic fault locators and can result in a fault location error [51]. Manual analysis may be required in such cases in which a user visually inspects both voltage and current, determines the correct fault type, and uses the correct equation to calculate fault location. Alternatively, the user may use a two-ended algorithm that does not require knowledge of the fault type (synchronized, unsynchronized, or unsynchronized negative-sequence methods) if data from both ends of the transmission line are available.

4.3   Application Errors

Impedance-based fault location algorithms described in Chapter 3 make certain simplifying assumptions when calculating the distance to fault. Accuracy is compromised when the algorithms are applied to systems in which those assumptions do not hold true. Examples include non-homogeneous systems or systems with significant load, parallel lines, series compensation, three-terminal lines, and tapped radial lines. The nature of the fault itself may further complicate the task of accurate fault location. This section evaluates the error in fault location due to such application-related challenges.

4.3.1   Load

We used the power system model to evaluate the impact of load on the accuracy of one- and two-ended impedance-based fault location algorithms. We simulated single line-to-ground faults at several locations of the 18-mile long transmission line and varied δ and R_f. Recall that δ represents the net load served by the transmission network, and R_f is the resistance of the fault. Single-ended fault location algorithms used voltage and current phasors captured at terminal G while two-ended algorithms used voltage and current phasors recorded at both terminals.

The results of the simple reactance method are shown in Figure 4.13. In the reference case, the transmission line was heavily loaded. A power angle of 20° translated to a load current of 430 A. Despite the presence of significant load current, the zero resistance of the fault allowed the simple reactance method to be 100 percent accurate. For non-zero values of fault resistance, however, the same load current caused a reactance error in the simple reactance method. The reactance error was capacitive and caused the simple reactance method to underestimate the fault location. The fault location error was further magnified

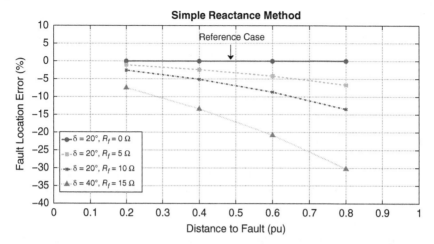

Figure 4.13 The accuracy of the simple reactance method deteriorates due to the combination of load and fault resistance.

when the load and fault resistance were increased to 40° and 15 Ω, respectively. Also notice that the reactance error increased as the fault moved farther away from terminal G. When a fault occurs toward the end of a transmission line, the fault current magnitude decreases. The load current makes up a more significant percent of the total fault current and increases the reactance error.

The Takagi method uses the pure fault current to minimize the reactance error due to load. As shown in Figure 4.14, the reactance error was negligible when $R_f = 10\,\Omega$ and $\delta = 20°$. The modified Takagi, the current distribution factor, and the two-ended methods were also not affected by an increase in the system load.

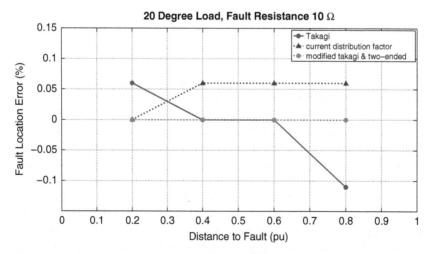

Figure 4.14 The accuracy of Takagi, modified Takagi, current distribution factor method, and two-ended methods was unaffected by load and fault resistance.

4.3.2 Non-Homogeneous System

We used the power system model to study the impact of a non-homogeneous system on impedance-based fault location algorithms. We started off with a reference case in which no changes were made to the existing power system model; a homogeneous system where the local and remote source impedances have the same phase angle as the line impedance. We then simulated single line-to-ground faults along the entire length of the transmission line with $\delta = 1°$ (very lightly loaded line) and $R_f = 5\,\Omega$. To compute the location of faults, single-ended methods used voltage and current waveforms at terminal G while two-ended methods used voltage and current measurements at both terminals. Next, we intentionally made the system non-homogeneous by changing $Z_{G_1}^{eq}$ to $15\angle50°\,\Omega$. We simulated faults using the same values of fault resistance and load. Location estimates from the one- and two-ended methods, computed using the new set of voltage and current phasors, were compared against those obtained in the reference case (homogeneous system). The results are displayed graphically in Figures 4.15, 4.16, 4.17, and 4.18. As expected, the accuracy of simple reactance and Takagi methods deteriorate in a non-homogeneous system. The current distribution factor method uses the remote source impedance to improve upon the performance of the Takagi method. The modified Takagi after angle correction and two-ended methods are not affected by the increase in system non-homogeneity.

4.3.3 Zero-Sequence Mutual Coupling

Single-circuit transmission lines shown in Figure 4.19 are the simplest means to transfer power from one area to another. Utilities will often parallel transmission lines or install series capacitors (discussed in the next section) to increase the power transfer capability of transmission lines. To better understand this, consider the power transfer equation given by:

$$P = \frac{V_1 V_2 \, sin\,(\theta_{12})}{X_L}, \tag{4.9}$$

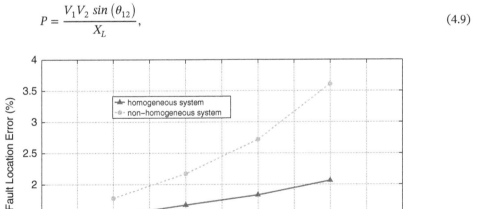

Figure 4.15 Impact of a non-homogeneous system on the simple reactance method.

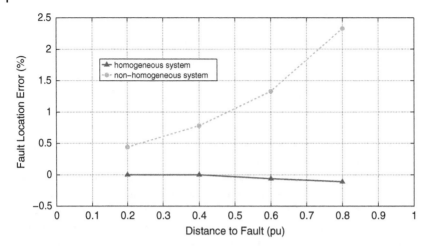

Figure 4.16 Impact of a non-homogeneous system on the Takagi method.

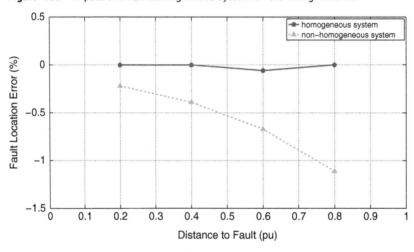

Figure 4.17 Impact of a non-homogeneous system on the current distribution factor method.

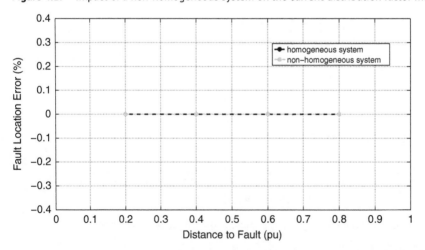

Figure 4.18 Modified Takagi (with angle correction) and two-ended methods unaffected by a non-homogeneous system.

Figure 4.19 Single-circuit transmission line.

where P is the power transferred from terminal 1 to terminal 2, V_1 is the voltage magnitude at terminal 1, V_2 is the voltage magnitude at terminal 2, θ_{12} is the angular difference between voltages at terminals 1 and 2, and X_L is the inductive reactance of the transmission line. We ignore the resistance of the transmission line in this equation. Now, voltage magnitudes V_1 and V_2 are fixed at 95 to 105 percent of the nominal voltage, and θ_{12} depends on system conditions. So the only variable that system planning engineers can manipulate to transfer more power is X_L. One option is to build additional transmission lines in parallel with the existing line. This reduces the effective X_L between two terminals. The other option is to install series capacitors. The capacitive reactance cancels part of the inductive reactance and decreases X_L. While both options increase power transfer through a transmission line, they introduce additional challenges to impedance-based fault location algorithms. In this section, we discuss the impact of parallel lines on fault location.

Figure 4.20 shows a multiple-circuit transmission line in which two three-phase lines are supported by the same tower. Alternatively, the two three-phase lines may run on two separate towers but be in close proximity to one another as shown in Figure 4.21. The close proximity causes the parallel lines to become magnetically coupled with one another. The coupled lines may be at the same or different voltage levels. The coupling mainly impacts the zero-sequence network and can impact the impedance to fault calculations during a ground fault [84].

Figure 4.20 Two three-phase lines supported by the same tower.

To understand the effects of mutual coupling, consider the 69 kV double-circuit transmission network shown in Figure 4.22. Source impedance is the same as that used to build the power system model in Section 4.1. The transmission line is 18 miles long and has the configuration shown in Figure 4.23. Phase conductors A, B, and C represent Line A while phase conductors A', B', and C' represent Line B. Conductor size and type are the same as those described in Table 4.1. Assuming both lines to be completely transposed and using an

Figure 4.21 The tower on the right has one three-phase circuit. However, due to its proximity to the other tower, it will be affected by mutual coupling.

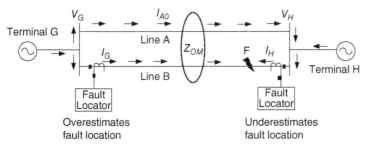

Figure 4.22 Double-circuit transmission network.

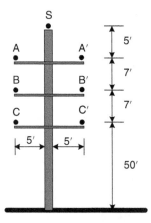

Figure 4.23 Configuration of a 69 kV double-circuit transmission line.

earth resistivity value of 100 Ω-m, the sequence impedance matrix Z_{012} of the transmission line is given by

$$Z_{012} = \begin{bmatrix} 15.82 + j40.91 & 0 & 0 & 10.52 + j25.77 & 0 & 0 \\ 0 & 5.33 + j12.92 & 0 & 0 & 0 & 0 \\ 0 & 0 & 5.33 + j12.92 & 0 & 0 & 0 \\ 10.52 + j25.77 & 0 & 0 & 15.82 + j40.91 & 0 & 0 \\ 0 & 0 & 0 & 0 & 5.33 + j12.92 & 0 \\ 0 & 0 & 0 & 0 & 0 & 5.33 + j12.92 \end{bmatrix} \Omega.$$

Here, the off-diagonal term $10.52 + j25.77 \, \Omega$ represents the zero-sequence mutual coupling (Z_{0M}) between two parallel lines and will always be present regardless of whether the line is transposed or not. Notice that Z_{0M} is significant, around 63 percent of the zero-sequence line impedance. This value is known as the coupling strength [85] and depends on the proximity between two lines and the distance for which the two lines are coupled. Two lines can be coupled the entire way as in our example system. They can also be coupled only up to a certain distance particularly if they start parallel to one other from one terminal but end at two different terminals. Mutual coupling can be neglected if the coupling strength is less than 10 percent.

In our example case, the coupling strength is significant and will affect the apparent impedance measured by the fault locators. Suppose that a bolted single line-to-ground fault occurs on Line B. The fault is very close to terminal H. As a result, fault current contributed by terminal G has two paths to reach the fault, Line A and Line B as shown in Figure 4.22. This ground current that is flowing in Line B, I_{A0}, induces a zero-sequence voltage in the other line due to mutual coupling and modifies the apparent impedance measured by the fault locator at terminal G as:

$$Z_{app} = \frac{V_G}{I_G} = mZ_1 + mZ_{0M} \left(\frac{I_{A0}}{I_G} \right). \tag{4.10}$$

The first term in the above equation is the actual impedance to fault while the second term is the additional voltage drop due to mutual coupling. Because both currents are flowing in the same direction, the fault locator will measure a higher impedance to fault and will overestimate the fault location. The opposite happens at terminal H. Because the fault is so close to that terminal, all the fault current contributed by the source at terminal H flows directly to the fault. On Line A, however, the current contributed by terminal G is in a direction opposite to the direction of the current measured by the fault locator at terminal H. As a result, mutual coupling will cause the fault locator to underestimate the distance to the fault. In other words, if the current on the faulted line flows in the same direction as the current on the parallel line, then single-ended fault locating techniques will overestimate the location of the fault. If the currents flow in opposite directions, single-ended methods will underestimate the location of the fault [36]. Some references suggest measuring I_{A0} and inputting the value to (4.10) to improve fault location accuracy. This will help only if the two lines are parallel to each other for the entire line length.

We evaluated the impact of Z_{0M} on impedance-based fault location algorithms with the 69 kV double-circuit transmission network. We began the analysis by first developing a reference case where there was no zero-sequence mutual coupling between the two lines. In reality, this is possible only when the two parallel lines are far apart from each other. We simulated bolted single line-to-ground faults at various locations along Line 2. The load angle was kept at $\delta = 10°$. To compute the location of faults, single-ended methods used the voltage and current waveforms at terminal G while two-ended methods used waveforms at both ends of the line. Next, we intentionally introduced Z_{0M} to the model and simulated the same faults. We compared the location estimates computed using the new waveforms with those obtained in the reference case. All single-ended methods were equally affected by Z_{0M} as shown in Figure 4.24. The error increased as the fault moved farther away from the terminal.

Among the two-ended methods, only the synchronized line current differential method was affected by mutual coupling. The two-ended synchronized or the two-ended

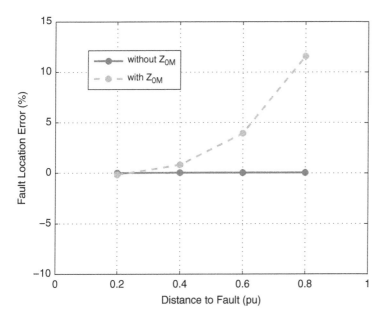

Figure 4.24 Mutual coupling affects accuracy of all single-ended methods.

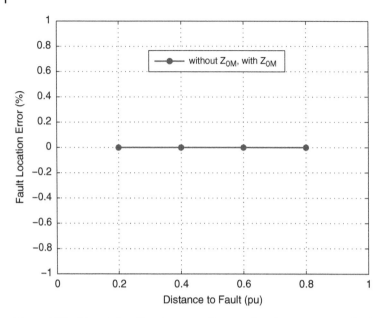

Figure 4.25 Mutual Coupling does not affect synchronized, unsynchronized, and unsynchronized negative-sequence two-ended methods.

unsynchronized methods do not use the zero-sequence network when computing the distance to a fault and were not affected by Z_{0M} as shown in Figure 4.25.

4.3.4 Series Compensation

Series compensation is another technique used by utilities to increase the power transfer capability of transmission lines as discussed in the previous section. Series capacitors can be installed at line terminals, either at one or both terminals, or in the middle of the transmission line. Line terminals are preferred due to space available inside the substation, which helps reduce installation costs [86–88]. Because series capacitors can experience a significant overvoltage during a fault, they are equipped with a spark gap or a metal oxide varistor (MOV) in parallel as shown in Figure 4.26. When voltage exceeds a certain threshold, the spark gap flashes over and removes the series capacitor from the circuit. The behavior of the MOV-protected series capacitor is different. Under normal conditions, the MOV is practically an open circuit (similar to the spark gap). However, when the voltage across the series capacitor exceeds a reference voltage, the MOV starts conducting and behaves as a nonlinear resistor, the value of the resistor being inversely proportional to the voltage. The MOV operation reduces the effective capacitance but does not remove the series capacitor completely from the circuit like the spark gap.

While there are many benefits of installing series capacitors, they pose a challenge to all impedance-based fault location algorithms. If the spark gap flashes over during a fault and shorts out the series capacitor, there will be no fault location error. However, if the spark gap does not flash over for low magnitude faults or the series capacitor is protected by an

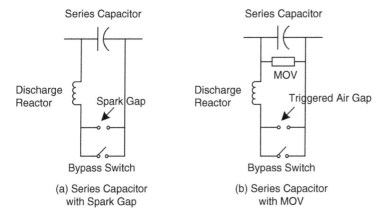

Figure 4.26 Series capacitor equipped with spark gap or MOV [86].

MOV, the apparent impedance seen by the fault locator will include the reactance of the series capacitors and can result in a significant error. If series capacitors are installed at the line terminal, perform fault location with line-side VTs if available. Avoid using bus-side VTs. Because impedance is measured from the location of the VT, using VT measurements beyond the series capacitor removes the capacitive reactance from the apparent impedance calculation and negates any fault location error.

4.3.5 Three-Terminal Lines

Single-ended impedance-based fault location algorithms described in Chapter 3 have been primarily developed for radial and two-terminal transmission lines. The accuracy of these methods get negatively affected when applied to locate faults on a three-terminal line. The algorithms are accurate only up to the tap point. When a fault occurs beyond the tap point, the fault current contributed by the third terminal (terminal T) modifies the impedance to fault equation and results in a significant error in location estimates. For example, consider the fault shown in Figure 4.27. This fault is beyond the tap point for terminal G. The apparent impedance measured from terminal G is:

$$Z_{app} = \frac{V_G}{I_G} = mZ_{L1} + (m - D) Z_{L1} \frac{I_T}{I_G} + R_f \left(\frac{I_F}{I_G} \right), \qquad (4.11)$$

where I_T is the fault current contributed by terminal T, and D is the distance of the tap point from terminal G. Since single-ended algorithms at terminal G have no knowledge about I_T, the term $(m - D) Z_{L1} (I_T/I_G)$ will cause single-ended methods to overestimate the location of the fault. In other words, the fault will appear to be farther away than it actually is. Single-ended methods applied from terminal H, on the other hand, can avoid this source of error when locating fault F. Since the fault is located before the tap point, the fault current contributed by terminals G and T act as remote infeed only.

The above discussion hints at a possible solution. When a fault occurs on a three-terminal line, apply single-ended methods from each terminal. Choose the estimate that is shorter

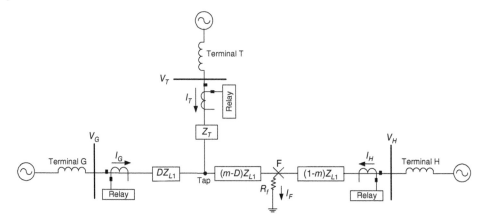

Figure 4.27 Three-terminal transmission line.

than the distance between the specific terminal and the tap point [54]. The other alternative is to apply the three-ended algorithms described in Chapter 3.

4.3.6 Radial Tap

In a three-terminal line, one of the terminals may be connected to a generating station. If the generating station goes offline, the line that connects it to the other two terminals becomes radial. Essentially, the three-terminal line becomes a two-terminal line with a tapped radial line. Utilities can also tap a long line from the main two-terminal line to serve a remote load [59]. Even though the tap line is radial, the challenges faced by fault locators that use the one- or the two-ended algorithms are the same as those experienced for a three-terminal line. For example, when a bolted fault occurs in the line section between the tap point and the load as illustrated in Figure 4.28, the apparent impedance seen from terminal G is given by:

$$Z_{app} = \frac{V_G}{I_G} = Z_{1-seg1} + mZ_{1-seg2} + mZ_{1-seg2} \times \frac{I_H}{I_G}. \tag{4.12}$$

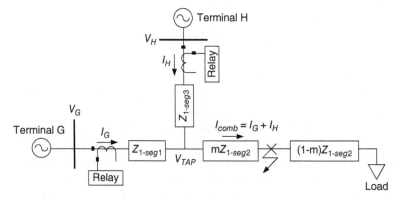

Figure 4.28 Radial line tapped from a two-terminal line.

The actual impedance to fault is the sum of the first two terms of the above equation. However, because both sources operate in parallel to contribute current to the fault, the impedance seen from terminal G includes an additional term, i.e., the third term in the above equation. Because single-ended algorithms make use of measurements captured at only one end of the line, neglecting the fault current contributed by terminal H will cause single-ended methods to always overestimate the distance to the fault location. Two-ended methods will also yield an erroneous estimate as the methods have been derived for a two-terminal line with no taps.

It is possible to do a manual analysis with measurements captured at both line ends to improve the accuracy of location estimates. The first step is to identify whether the fault has occurred on the line between two terminals or on the radial tap line. To do this, calculate the voltage at the tap point during fault, V_{Tap}, from terminals G and H as shown below:

$$\text{Terminal G:} \quad V_{Tap} = V_G - Z_{1-seg1}I_G, \tag{4.13}$$

$$\text{Terminal H:} \quad V_{Tap} = V_H - Z_{1-seg3}I_H. \tag{4.14}$$

If the fault is on the radial line, V_{Tap} calculated from terminal G will equal that calculated from terminal H. This is because terminals G and H operate in parallel to feed any fault beyond the tap point. Next, apply single-ended impedance-based fault location algorithms using V_{Tap} and I_{comb} (summation of both terminal currents) to compute the distance to fault from the tap.

4.3.7 Evolving Faults

Impedance-based fault location algorithms expect fault location to be calculated when the fault has reached steady state. Unfortunately, faults are not always well behaved. Many faults start out as single line-to-ground faults but then evolve to include the other phases as well. In some cases, the fault type may remain the same, but the fault resistance may change drastically during the duration of the fault. For example, Figure 4.29 shows a line-to-line

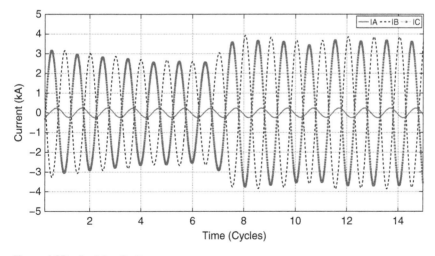

Figure 4.29 Evolving fault.

fault that began as a high impedance fault but evolved into a bolted fault at around the seventh cycle, as indicated by the increased fault current magnitude. Sequential clearing of faults on a transmission line can also make a fault appear to be dynamic. For example, in a two-terminal line, the station closest to the fault may trip instantaneously on Zone 1. The station at the other end of the line may see the fault in Zone 2 and trip after an intentional time delay. When the station closest to the fault trips, the other end will see an increase in fault current contribution. Such faults that evolve over time may pose a challenge to automatic fault locators. If they happen to calculate fault location right when the fault evolves, there may be a significant error in the location estimate. An offline analysis is necessary to choose the best cycle for determining the fault location.

4.4 Exercise Problems

■ Exercise 4.1

A 34.5 kV distribution feeder experienced a single line-to-ground fault on phase A. The distribution feeder has a positive- and zero-sequence impedance of $z_1 = 0.11 + j0.60$ Ω/mile and $z_0 = 0.39 + j2.62$ Ω/mile, respectively. An overcurrent relay at the substation captured an 11-cycle snapshot of the voltages and currents during the fault. The waveforms are shown in Figure 4.30 and a circuit model of the power system is shown in Figure 4.31. Notice that the relay is measuring phase-to-phase voltages. The phase current and phase-to-phase voltage phasors during the fault are given below:

$$I^f_{G_{abc}} \text{ [kA]} \qquad V^f_{G_{ab-bc-ca}} \text{ [kV]}$$
$$\begin{bmatrix} 2.21\angle113.25° \\ 0.18\angle -50.32° \\ 0.21\angle82.77° \end{bmatrix} \qquad \begin{bmatrix} 34.51\angle153.83° \\ 35.89\angle -84.86° \\ 36.43\angle36.43° \end{bmatrix}.$$

The zero-sequence source impedance is $Z^{eq}_{G_0} = 0.02 + j2.97$ Ω. Given the above data, you have been tasked to calculate the fault location and provide that information to the field crew. Use the simple reactance method.

Solution:

The zero-sequence fault current during the ground fault can be calculated as:

$$3I^f_{G_0} = I^f_{G_a} + I^f_{G_b} + I^f_{G_c}$$
$$= 2.21\angle113.25° + 0.18\angle -50.32° + 0.21\angle82.77°$$
$$= 2.22\angle109.19° \quad \text{[kA]}.$$

Next, let us calculate the zero-sequence compensation factor as:

$$k = \frac{z_0 - z_1}{z_1}$$
$$= \frac{(0.39 + j2.62) - (0.11 + j0.60)}{0.11 + j0.60}$$
$$= 3.34\angle2.50°.$$

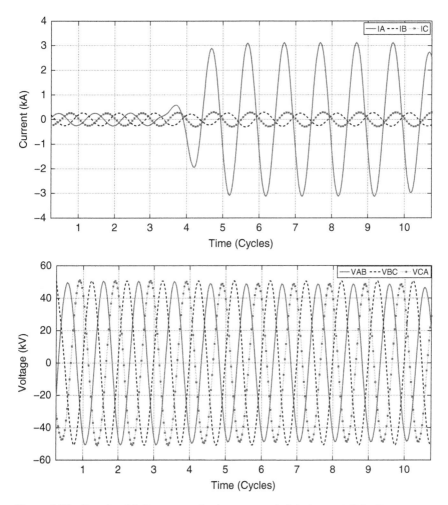

Figure 4.30 Exercise 4.1: Currents and voltages recorded during an A-G fault.

Current I_G during an A-G fault is:

$$I_G = I_{G_a}^f + k \times I_{G_0}^f$$

$$= 2.21\angle113.25° + \frac{3.34\angle2.50° \times 2.22\angle109.19°}{3}$$

$$= 4.68\angle112.43° \quad [\text{kA}].$$

Because single-ended impedance-based fault location algorithms require the input of line-to-ground voltages when calculating the distance to a single line-to-ground fault, the next task is to estimate the faulted phase voltage using (4.6):

$$V_{G_{a-est}}^f = \frac{1}{3}\left(V_{G_{ab}}^f - V_{G_{ca}}^f\right) - Z_{G_0}^{eq} I_{G_0}^f$$

$$= \frac{1}{3}\left[(34.51\angle153.83° - 36.43\angle36.43°) - (0.02 + j2.97) \times 2.22\angle109.19°\right]$$

$$= 18.07\angle-175.46° \quad [\text{kV}].$$

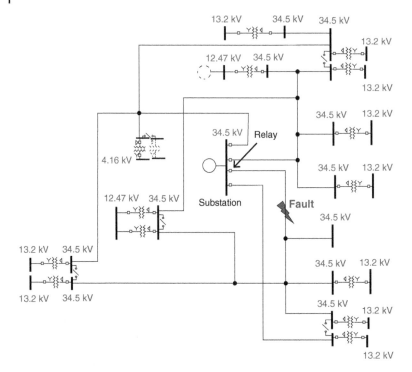

Figure 4.31 Exercise 4.1: One line.

Using the estimated faulted phase voltage, the distance to fault is calculated using the simple reactance method:

$$m = \frac{\text{imag}\left(\frac{V_G}{I_G}\right)}{\text{imag}\left(z_1\right)}$$

$$= \frac{\text{imag}\left(\dfrac{18.07\angle - 175.46°}{4.68\angle 112.43°}\right)}{\text{imag}\left(0.11 + j0.60\right)}$$

$$= 6.12 \quad [\text{mi}].$$

The short-circuit model was used to confirm the accuracy of the location estimate. An A-G fault was simulated at 6.12 miles from the substation. The model indicated a short-circuit current of 2.29 kA at the substation, which was close to the actual fault current, 2.21 kA, measured by the relay. This example illustrates the procedure for locating single line-to-ground faults with phase-to-phase voltages.

■ Exercise 4.2

At 2:00 a.m. on a Sunday, you get a call that a 345 kV transmission line running between terminal G and terminal H has experienced a fault. The relays protecting the transmission line went through a one-shot reclose sequence and locked out, indicating that the line

Figure 4.32 Exercise 4.2: One line.

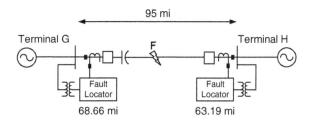

experienced a permanent fault. The line is 95 miles long. A fault locator at terminal G that was using the Takagi method indicated that the fault was at 68.66 miles from its terminal as shown in Figure 4.32. A fault locator at terminal H that was also using the Takagi method indicated that the fault was at 63.19 miles from its terminal. Clearly the results from both fault locators were not adding up to the total line length. For example, if the fault locator at one end indicated that the fault was at 68.66 miles, then the fault locator at the other end should have indicated that the fault was at 26.34 miles. Instead, the other end indicated 63.19 miles. The night crew needs your help to figure out where to go and search for the fault. This is a critical line and it is essential to find the fault and re-energize the line as quickly as possible. Given the above data, answer the following questions:

(a) Where should the line crew go and search for the fault? Why?
(b) Someone on the team remembered that at a training class, they had learned that two-ended methods are more accurate that single-ended methods. The team member suggests that before they head out to the field, they should gather events from both fault locators and run a double-ended fault location calculation manually. The team member asks for your thoughts. What would you tell them?

Solution:

(a) The line crew should search for the fault at 63.19 miles from terminal H. In other words, the fault locator at terminal H is correct. The data from the fault locator at terminal G should not be trusted as this fault locator is using bus-side potential, and hence its impedance to fault calculation is going to be affected by the presence of the series capacitor.
(b) It is true that two-ended methods are more accurate than single-ended methods. However, in this specific case where there is a series capacitor at one end of the line, two-ended methods would be inaccurate and should not be used.

Using your guidelines, the night crew found the fault at a tower that was 65 miles from terminal H.

■ Exercise 4.3

Efforts are underway to modernize the substations at terminal G and terminal H. Management would like to prevent a recurrence of the situation described in Exercise 4.2. They call you and ask if there is a technology that would allow them to locate faults on a series-compensated line from either terminal. What would you tell them?

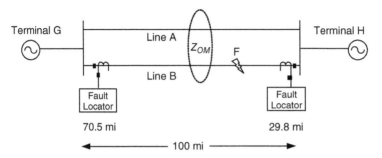

Figure 4.33 Exercise 4.4: One line.

Solution:

They can consider using fault locators that use the traveling-wave technology. Series capacitors allow the high frequency traveling waves to pass through, appearing as an effective short circuit to these waves. As a result, series capacitors do not affect the accuracy of fault locators that use the traveling-wave technology.

■ Exercise 4.4

During a severe thunderstorm, there was a line-to-line fault on Line B (see Figure 4.33), causing the protective relaying for that line to trip and isolate the fault. Line B is 100 miles long. The fault locator at terminal G indicated the fault location to be at 70.5 miles from its end while the fault locator at terminal H indicated the fault to be at 29.8 miles from its end. Your colleagues are concerned about the proximity of Line A to Line B. They had heard about mutual coupling and are wondering whether they can trust the fault location result from the fault locators. The fault locators use the Takagi method to calculate fault location. What would you tell them?

Solution:

This was a trick question! Your colleagues had a valid concern. But after reading through an awesome book on fault location, you know that mutual coupling affects the Takagi method during a single line-to-ground fault. Because this is a line-to-line fault, the result from the fault locator should be accurate. Plus the results from both fault locators add up to the line length, which is another sign that you can trust the fault locator result. You convey this information to your colleagues.

4.5 Summary

Table 4.4 summarizes the sources of fault location error evaluated in this chapter and their corresponding impact on impedance-based fault location algorithms.

Table 4.4 Summary of Fault-Locating Error Sources That Affect Impedance-Based Fault Location Algorithms

Input Data	Simple Reactance	Takagi	Modified Takagi (angle correction)	Current Distribution Factor	Current Differential Two-ended	Synchronized Two-ended	Unsynchronized Two-ended	Unsynchronized Negative-Sequence Two-ended
Input Data Errors								
DC offset	✓	✓	✓	✓	✓	✓	✓	✓
CT saturation	✓	✓	✓	✓	✓	✓	✓	✓
Aging CCVTs	✓	✓	✓	✓	✓	✓	✓	✓
Open-delta PT[1]	✓	✓	✓	✓	✓			
Line length	✓	✓	✓	✓	✓	✓	✓	✓
Untransposed lines	✓	✓	✓	✓	✓	✓	✓	✓
Earth resistivity[2]	✓	✓	✓	✓	✓	✓		
Non-homogeneous lines	✓	✓	✓	✓	✓	✓	✓	✓
Fault type	✓	✓	✓	✓	✓			
Application Errors								
Load	✓							
Non-homogeneous system	✓	✓						
Mutual coupling[3]	✓	✓	✓	✓	✓			
Series compensation	✓	✓	✓	✓	✓	✓		
Three-terminal[4] lines	✓	✓	✓	✓	✓	✓	✓	✓
Tapped radial[4] lines	✓	✓	✓	✓	✓	✓	✓	✓
Evolving fault	✓	✓	✓	✓	✓	✓	✓	✓

1) Open-delta PTs pose a problem in locating single line-to-ground faults only. If the zero-sequence impedance of the local source is available, estimate the corresponding line-to-ground voltages.
2) Earth resistivity affects the accuracy of locating single line-to-ground faults only.
3) Mutual coupling affects the accuracy of single-ended algorithms in locating single line-to-ground faults only. If transmission lines are parallel for the entire line length, then the residual current from the parallel line can improve the accuracy of single-ended methods.
4) It is possible to modify two-ended methods for application to tapped lines and three-terminal lines.

5

Fault Location on Overhead Distribution Feeders

A distribution system is an important part of the power system infrastructure. They can be seen everyday along the side of roads and streets. They take electricity from transmission and sub-transmission lines with voltages 69 kV and above and deliver it to residential customers, business entities, or industrial facilities. Anything below 35 kV is normally considered to be distribution. Some standard primary distribution voltages are 4.16 kV, 12.47 kV, 13.2 kV, 14.4 kV, 23.9 kV, and 34.5 kV [89].

A transformer at a distribution substation steps down the voltage of incoming transmission lines to one of the standard primary distribution voltages. The size of the transformer depends on the size of the load that is being served. In rural areas, the distribution transformer may be rated for 5 MVA while in densely populated areas, it may be rated for 200 MVA. Smaller substations may not necessarily be fed by transmission or sub-transmission lines. The incoming line can come from another distribution feeder that is at a higher voltage than the voltage of the substation. Outgoing distribution feeders then carry the electricity from the substation to a point close to the end user where another distribution transformer steps down the primary distribution voltage to a secondary distribution voltage, normally 120/240 V in the US. It is this secondary distribution circuit that serves end users.

Figure 5.1 shows the schematic of a typical distribution network. You can see that these networks are quite extensive. The feeders leaving the substation form the backbone of the circuit and are often referred to as the mains. They normally consist of three phase conductors and a neutral conductor that is grounded at multiple points along the length of the feeder. The phase conductors are typically designed to carry 400 A of continuous current and 600 A during emergency conditions for a short time interval [71]. The mains are protected by overcurrent relays at the substation. There should ideally be one relay per feeder. When a short-circuit fault occurs on the distribution feeder, the relay detects the fault and opens up the feeder circuit breaker at the substation.

Several laterals tap off from the mains to serve customers. They may be single-, two-, or three-phase branches with a neutral conductor that is grounded at multiple points. Figure 5.2 shows a three-phase lateral with two single-phase load taps. A fuse protects each tap from overcurrent condition due to faults. A lightning arrester protects the tap from any overvoltage condition. Each tap feeds an underground cable that is serving residential customers through a pad-mounted transformer (not shown in the figure). Figure 5.3 shows

Fault Location on Transmission and Distribution Lines: Principles and Applications, First Edition.
Swagata Das, Surya Santoso, and Sundaravaradan N. Ananthan.
© 2022 John Wiley & Sons Ltd. Published 2022 by John Wiley & Sons Ltd.
Companion website: www.wiley.com/go/das/faultlocation

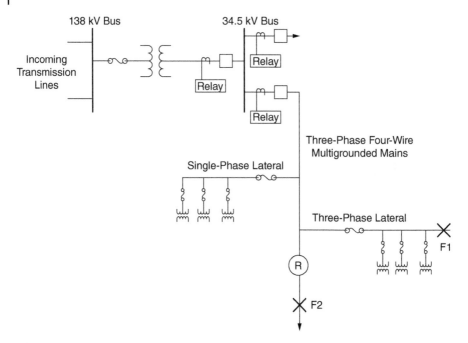

Figure 5.1 Typical schematic of a distribution substation. Protective devices shown only for the 34.5 kV bus and downstream feeders and laterals.

a three-phase feeder with a single-phase taps while Figure 5.4 shows a three-phase feeder with three-phase taps. Three-phase taps are utilized when serving larger loads.

Utilities may implement either a fuse tripping or fuse saving scheme in their distribution circuit. For example, consider the fault shown as F1 in Figure 5.1. The utility may design the protection system such that the fuse protecting the lateral blows before the relay at the substation can trip the circuit breaker, thus isolating the faulted lateral from the mains. This is known as a fuse tripping scheme. In such a scheme, customers on other healthy laterals are not affected by the fault. They only see a momentary voltage sag during the duration of the fault. Voltage is subsequently restored back to nominal after the fuse operation. The scheme, however, costs the utility time, money, and resources when the fuse blows unnecessarily during a temporary fault. Line crew need to be dispatched at short notice to replace the blown fuse.

The fuse-saving scheme, as the name suggests, attempts to save the fuse during a temporary fault. The protection scheme is designed such that the relay at the substation trips before the fuse can melt for a fault at F1. The circuit breaker stays open for a certain time interval, known as the open interval, during which the fault has sufficient time to clear out on its own. After the open interval times out, the relay issues a close signal to the circuit breaker. If the fault is still present, the relay at the substation may make several more attempts to clear the temporary fault. The number of attempts is programmable by the utility. Once all attempts to clear the fault fails, it is deemed a permanent fault. The relay switches to the delay curve, which allows the fuse to blow first and isolate the permanent fault. The disadvantage of such a scheme is that all customers, including those on healthy laterals, experience momentary interruptions.

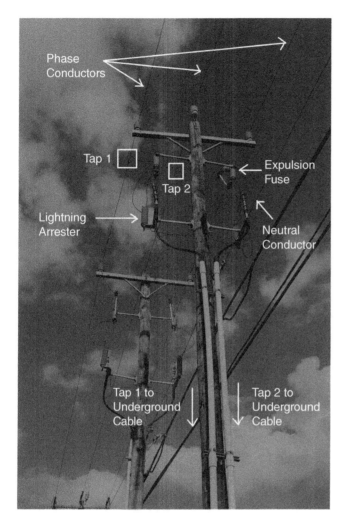

Figure 5.2 Three-phase lateral with two load taps. The load taps are descending down the pole and going underground to serve customers through a pad-mounted transformer.

Additional reclosers may be installed along the mains to improve selectivity. These reclosers sit on top of distribution poles as shown in Figure 5.5. They are controlled by a recloser controller that sits at the bottom of the pole as shown in Figure 5.6 and is powered from the line. When there is a fault downstream from the recloser, F2 in Figure 5.1, the recloser clears the fault before the relay at the substation. Only customers downstream from the recloser are affected, thus improving selectivity.

Distribution feeders may be overhead or underground. Urban areas mostly consist of underground distribution feeders. In rural areas, feeders are typically overhead while in suburban areas, there is a good mix of overhead and underground feeders. Underground construction is more expensive than overhead construction but improves overall aesthetics. There is no distribution pole every 150 feet and all power lines are buried underground.

Figure 5.3 Three-phase lateral with a single load tap.

Underground feeders are more reliable as the buried power lines are not exposed to nature, animals, or trees. Overhead lines, on the other hand, experience a number of temporary faults due to lightning and contact with animals or trees. Underground construction also reduces maintenance costs as there is no need to set up a maintenance schedule to trim vegetation. Underground construction increases public safety as there is less likelihood of someone making accidental contact with energized equipment. However, should a fault occur on an underground cable, it takes a longer time to find the fault and perform repairs.

Many different methods are used by utilities to locate faults on underground cables [90]. One approach is the trial and error method. On the radial tap with the blown fuse, the cable is opened up in the middle and the fuse is replaced. If the fuse blows, the fault must be upstream from the center of the cable. If the fuse does not blow, then the fault must be downstream from the center of the cable. Once the faulted section is identified, the same process is repeated again by sectionalizing the faulted section in the middle and replacing the fuse until the crew is able to narrow down the fault location. A second method is to use

Figure 5.4 Three-phase lateral with a three-phase load tap going underground to serve a large residential complex.

faulted circuit indicators made primarily for underground systems. These indicators clamp around the conductor and notify the crew when they measure fault current. Another method known as the thumper can be used to precisely locate the cable fault so that the crew can start digging into the ground. It sends a pulsed DC signal through the cable. The repeated spark-over at the fault point generates a thumping noise, which enables the crew to locate the fault. Another method known as time domain reflectometry sends a very short-duration current pulse into the cable. At discontinuities such as a fault, the signal will get reflected back. Knowing the propagation velocity of the current pulse through the cable, it is easy to estimate the distance to the fault. Notice that we didn't mention using impedance-based fault location methods that we learned in Chapter 3 to locate faults on underground cables. The capacitance in underground cables is significant, which affects the value of the zero-sequence impedance. This makes it very difficult to locate faults using impedance-based methods.

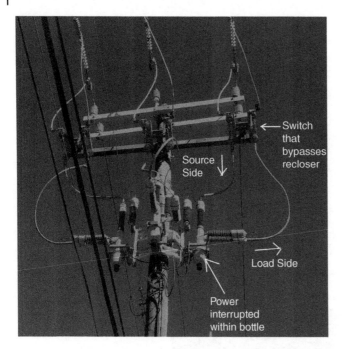

Source
Side

← Switch
that
bypasses
recloser

→
Load Side

Power
interrupted
within bottle

Figure 5.5 Recloser mounted on distribution poles improves selectivity.

In this chapter, we restrict our discussion to locating faults on overhead distribution feeders. We review the principles, strengths, and weaknesses of different fault-locating techniques proposed in the literature, discuss the challenges associated with locating faults on overhead distribution feeders, and solve the distance to fault equations with field data available from actual distribution faults.

5.1 Impedance-Based Methods

The one-ended impedance-based fault location algorithms described in Chapter 3 can be used to locate faults on a distribution feeder as well. These algorithms use voltage and current recorded by a protective relay or an intelligent electronic device during a fault to estimate the impedance to the fault; hence the name impedance-based methods. Some use prefault data to improve the accuracy of the impedance estimate. While the impedance to fault information is useful, line patrol prefer to get the same information but in units of distance rather than in ohms. As a result, fault location algorithms require the user to enter the line impedance per unit mile. This additional information allows the fault location algorithms to convert the estimated impedance to fault to a distance estimate. The protective relay or the intelligent electronic device can be located at the substation or at any other point along the distribution feeder upstream from the fault. In this section, we review traditional impedance-based fault location algorithms proposed in the literature for distribution systems.

Figure 5.6 Recloser controller mounted inside an enclosure.

5.1.1 Loop Reactance Method

The loop reactance method simplifies the fault location problem by assuming an unloaded distribution feeder and calculating the loop reactance to the fault. We describe the principle behind this algorithm by using a simple distribution feeder shown in Figure 5.7. We assume that the feeder is l miles long and has a positive- and zero-sequence line impedance of z_1 and z_0 ohms per mile, respectively. The transmission network upstream from the substation is represented by an ideal source with a positive- and zero-sequence impedance of Z_{src_1} ohms and Z_{src_0} ohms, respectively. A protective relay at the substation provides overcurrent protection to the feeder. Current transformers and voltage transformers bring in current and voltage measurements to the relay. Under normal load conditions, the relays measure the voltage and current phasors on the three phases, $V^p_{S_{abc}}$ and $I^p_{S_{abc}}$, respectively. The superscript p stands for prefault.

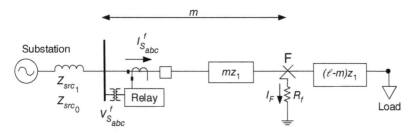

Figure 5.7 A simple distribution feeder experiencing a fault.

(1) Single Line-to-Ground Fault

Suppose that a single line-to-ground fault has occurred at a distance of m miles from the substation with a fault resistance of R_f ohms. The relay at the substation recorded the voltage and current phasors on the three phases during the fault, $V^f_{S_{abc}}$ and $I^f_{S_{abc}}$, respectively. The superscript f stands for fault. Since this was an unbalanced fault, we take the help of symmetrical components to estimate m.

The interconnection of the positive-, negative-, and zero-sequence networks during a single line-to-ground fault is shown in Figure 5.8. Note that we are assuming an unloaded

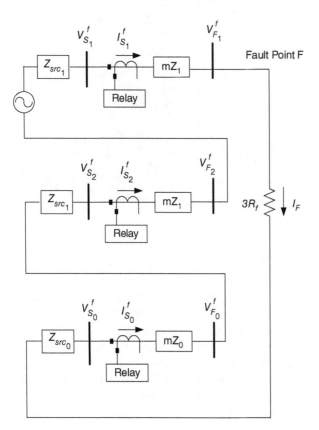

Figure 5.8 Interconnection of sequence networks during a single line-to-ground fault.

distribution feeder. The positive-sequence voltage phasor at the fault point, $V_{F_1}^f$, can be written as:

$$V_{F_1}^f = V_{S_1}^f - mz_1 \times I_{S_1}^f \quad [V],$$ (5.1)

where $V_{S_1}^f$ is the positive-sequence voltage phasor measured by the relay at the substation during the fault in volts and $I_{S_1}^f$ is the positive-sequence current phasor recorded by the relay at the substation during the fault in amperes. The negative-sequence voltage phasor at the fault point, $V_{F_2}^f$, can be written as:

$$V_{F_2}^f = V_{S_2}^f - mz_1 \times I_{S_2}^f \quad [V],$$ (5.2)

where $V_{S_2}^f$ is the negative-sequence voltage phasor measured by the relay at the substation during the fault in volts and $I_{S_2}^f$ is the negative-sequence current phasor recorded by the relay at the substation during the fault in amperes. The zero-sequence voltage phasor at the fault point, $V_{F_0}^f$, can be expressed as:

$$V_{F_0}^f = V_{S_0}^f - mz_0 \times I_{S_0}^f \quad [V],$$ (5.3)

where $V_{S_0}^f$ is the zero-sequence voltage phasor measured by the relay at the substation during the fault in volts and $I_{S_0}^f$ is the zero-sequence current phasor recorded by the relay at the substation during the fault in amperes. Adding (5.1), (5.2), and (5.3), we get the voltage phasor of the faulted phase at the fault point, $V_{F_a}^f$:

$$
\begin{aligned}
V_{F_a}^f &= V_{F_0}^f + V_{F_1}^f + V_{F_2}^f \\
&= V_{S_a}^f - mz_1 \left(I_{S_1}^f + I_{S_2}^f \right) - mz_0 I_{S_0}^f \quad [V].
\end{aligned}
$$ (5.4)

The voltage at the fault point, $V_{F_a}^f$, is nothing but the voltage drop across the fault resistance and can be written as:

$$V_{F_a}^f = 3I_{S_0}^f R_f \quad [V].$$ (5.5)

Combining the above two equations and re-arranging the terms, we get:

$$V_{S_a}^f = mz_1 \left(I_{S_1}^f + I_{S_2}^f \right) + mz_0 I_{S_0}^f + 3I_{S_0}^f R_f \quad [V].$$ (5.6)

The loop reactance method assumes that there is no load on the system. As a result, the sequence currents $I_{S_1}^f$, $I_{S_2}^f$, and $I_{S_0}^f$ are all equal to each other. This simplification allows us to write (5.6) as:

$$
\begin{aligned}
V_{S_a}^f &= mz_1 \left(I_{S_0}^f + I_{S_0}^f \right) + mz_0 I_{S_0}^f + 3I_{S_0}^f R_f \\
&= mI_{S_0}^f \left(2z_1 + z_0 \right) + 3I_{S_0}^f R_f \quad [V].
\end{aligned}
$$ (5.7)

Dividing throughout by $I_{S_0}^f$, we get

$$\frac{V_{S_a}^f}{I_{S_0}^f} = m\left(2z_1 + z_0\right) + 3R_f \quad [\Omega].$$ (5.8)

The above equation has 2 unknowns: m and R_f. Taking advantage of the fact that fault resistance is resistive in nature, the loop reactance method equates only the imaginary parts to get rid of fault resistance. The resulting expression for the distance to fault calculation is

$$m = \frac{\text{imag}\left(\dfrac{V_{S_a}^f}{I_{S_0}^f}\right)}{\text{imag}\left(2z_1 + z_0\right)} \quad \text{[mi]}. \tag{5.9}$$

Generalizing the above equation for all single line-to-ground faults, we get:

$$m = \frac{\text{imag}\left(\dfrac{V_{phase}}{I_0}\right)}{\text{imag}\left(2z_1 + z_0\right)} \quad \text{[mi]}, \tag{5.10}$$

where V_{phase} is the voltage of the faulted phase and I_0 is the measured zero-sequence current during the fault.

(2) Line-to-Line Fault

Suppose that a line-to-line fault occured at a distance of m miles from the substation with a fault resistance of R_f ohms. The loop reactance method makes use of symmetrical components to solve for m. The positive- and negative-sequence networks are connected in parallel at the fault point as shown in Figure 5.9. Again, note that we are assuming an unloaded feeder. The positive-sequence voltage phasor at the fault point, $V_{F_1}^f$, can be written as:

$$V_{F_1}^f = V_{S_1}^f - mz_1 \times I_{S_1}^f \quad \text{[V]}. \tag{5.11}$$

The negative-sequence voltage phasor at the fault point, $V_{F_2}^f$, can be written as:

$$V_{F_2}^f = V_{S_2}^f - mz_1 \times I_{S_2}^f \quad \text{[V]}. \tag{5.12}$$

Because the loop reactance method ignores load, $I_{S_2}^f = -I_{S_1}^f$ in the above equation. Also, for a line-to-line fault, the difference between $V_{F_1}^f$ and $V_{F_2}^f$ is nothing but the voltage drop across

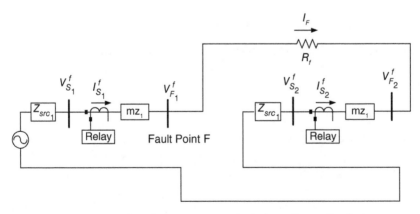

Figure 5.9 Interconnection of sequence networks during a line-to-line fault.

the fault resistance as shown below.

$$V_{F_1}^f - V_{F_2}^f = I_F R_f,$$

$$\therefore V_{S_1}^f - V_{S_2}^f - mz_1 \left(I_{S_1}^f + I_{S_1}^f \right) = I_{S_1}^f R_f \quad [\text{V}].$$

(5.13)

Dividing throughout by $I_{S_1}^f$, we get:

$$\frac{V_{S_1}^f - V_{S_2}^f}{I_{S_1}^f} - 2mz_1 = R_f \quad [\Omega].$$

(5.14)

Equating only the imaginary parts and rearranging the terms, we can solve for m as

$$m = \frac{\text{imag}\left(\dfrac{V_{S_1}^f - V_{S_2}^f}{I_{S_1}^f} \right)}{\text{imag}\left(2z_1 \right)} \quad [\text{mi}].$$

(5.15)

(3) Double Line-to-Ground Fault

During a double line-to-ground fault, the positive-, negative-, and zero-sequence networks are connected in parallel at the fault point as shown in Figure 5.10. We are assuming an unloaded distribution feeder. The positive-sequence voltage phasor at the fault point, $V_{F_1}^f$, can be written as:

$$V_{F_1}^f = V_{S_1}^f - mz_1 \times I_{S_1}^f \quad [\text{V}].$$

(5.16)

The negative-sequence voltage phasor at the fault point, $V_{F_2}^f$, can be written as:

$$V_{F_2}^f = V_{S_2}^f - mz_1 \times I_{S_2}^f \quad [\text{V}].$$

(5.17)

Because $V_{F_1}^f = V_{F_2}^f$ for a double line-to-ground fault, we equate the above two equations to arrive at the following expression:

$$V_{S_1}^f - V_{S_2}^f = mz_1 \left(I_{S_1}^f - I_{S_2}^f \right) \quad [\text{V}].$$

(5.18)

Dividing by $\left(I_{S_1}^f - I_{S_2}^f \right)$ and equating only the imaginary components, we get:

$$m = \frac{\text{imag}\left(\dfrac{V_{S_1}^f - V_{S_2}^f}{I_{S_1}^f - I_{S_2}^f} \right)}{\text{imag}\left(z_1 \right)} \quad [\text{mi}].$$

(5.19)

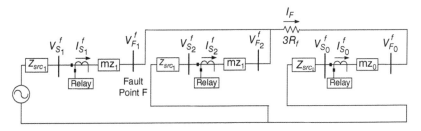

Figure 5.10 Interconnection of sequence networks during a double line-to-ground fault.

(4) Three-Phase Fault

A three-phase fault is a balanced fault and allows us to analyze any one of the three phases. In this derivation, we choose the A-phase. Similar to what we have been doing so far, we write the expression for the A-phase voltage at the faulted point as:

$$V_{F_a}^f = I_{F_a}^f R_f = V_{S_a}^f - mz_1 \times I_{S_a}^f \quad \text{[V]}. \tag{5.20}$$

Because the loop reactance method assumes a no-load condition, the fault current equals the current measured at the substation ($I_{F_a}^f = I_{S_a}^f$). Dividing throughout by $I_{S_a}^f$ and equating only the imaginary components, the distance to fault can be estimated as

$$m = \frac{\text{imag}\left(\dfrac{V_{S_a}^f}{I_{S_a}^f}\right)}{\text{imag}(z_1)} \quad \text{[mi]}. \tag{5.21}$$

Generalizing the above equation so that any phase may be used to calculate fault location, we get:

$$m = \frac{\text{imag}\left(\dfrac{V_{phase}}{I_{phase}}\right)}{\text{imag}(z_1)} \quad \text{[mi]}. \tag{5.22}$$

When all three phase currents are not exactly equal in magnitude, use the positive-sequence voltage and current to determine fault location as shown in the equation below:

$$m = \frac{\text{imag}\left(\dfrac{V_{S_1}^f}{I_{S_1}^f}\right)}{\text{imag}(z_1)} \quad \text{[mi]}. \tag{5.23}$$

5.1.2 Simple Reactance Method

The simple reactance method is another variation of the loop reactance method. Like the loop reactance method, this method also assumes that the distribution feeder has no load at the time of the fault. This method was derived in detail in Chapter 3. For brevity, we only list the equation in this chapter.

$$m = \frac{\text{imag}\left(\dfrac{V_S}{I_S}\right)}{\text{imag}(z_1)} \quad \text{[mi]}. \tag{5.24}$$

Here V_S and I_S depend on the fault type and are defined in Table 5.1.

5.1.3 Takagi Method

The Takagi method improves on the loop reactance and simple reactance methods by considering load current before the fault. To do this, it requires the user to enter the prefault current. The prefault current is then subtracted from the measured fault current

Table 5.1 Definition of V_S, I_S, ΔI_S, and I_{seq} for All Fault Types.

Fault Type	V_S	I_S	ΔI_S	I_{seq}
A-G	$V^f_{S_a}$	$I^f_{S_a} + kI^f_{S_0}$	$I^f_{S_a} - I^p_{S_a}$	$I^f_{S_2}$
B-G	$V^f_{S_b}$	$I^f_{S_b} + kI^f_{S_0}$	$I^f_{S_b} - I^p_{S_b}$	$I^f_{S_2}$
C-G	$V^f_{S_c}$	$I^f_{S_c} + kI^f_{S_0}$	$I^f_{S_c} - I^p_{S_c}$	$I^f_{S_2}$
AB or AB-G	$V^f_{S_a} - V^f_{S_b}$	$I^f_{S_a} - I^f_{S_b}$	$\left(I^f_{S_a} - I^p_{S_a}\right) - \left(I^f_{S_b} - I^p_{S_b}\right)$	$jI^f_{S_2}$
BC or BC-G	$V^f_{S_b} - V^f_{S_c}$	$I^f_{S_b} - I^f_{S_c}$	$\left(I^f_{S_b} - I^p_{S_b}\right) - \left(I^f_{S_c} - I^p_{S_c}\right)$	$jI^f_{S_2}$
CA or CA-G	$V^f_{S_c} - V^f_{S_a}$	$I^f_{S_c} - I^f_{S_a}$	$\left(I^f_{S_c} - I^p_{S_c}\right) - \left(I^f_{S_a} - I^p_{S_a}\right)$	$jI^f_{S_2}$
ABC	same as	line-to-line fault	or double line-to-ground fault	N/A

where $k = \dfrac{z_0 - z_1}{z_1}$

to calculate the pure fault that exists only during a fault, ΔI_S. The fault location is then calculated as shown below:

$$m = \frac{\text{imag}\left(V_S \times \Delta I_S^*\right)}{\text{imag}\left(z_1 \times I_S \times \Delta I_S^*\right)} \quad \text{[mi]}. \tag{5.25}$$

Here V_S, I_S, and ΔI_S depends on the fault type and is defined in Table 5.1 for every fault type. For detailed derivation of this method, refer to Chapter 3.

5.1.4 Modified Takagi Method

The modified Takagi method also accounts for load but does not require the user to enter any prefault data. It uses negative-sequence current instead of ΔI_S as pure fault current. From the chapter on symmetrical components, remember that the negative-sequence current does not exist during normal load conditions but materializes only during an unbalanced condition such as an unbalanced fault. Therefore, the negative-sequence current is technically a pure fault current and is free from any effects of load. The distance to fault is given by the equation below:

$$m = \frac{\text{imag}\left(V_S \times I_{seq}^*\right)}{\text{imag}\left(z_1 \times I_S \times I_{seq}^*\right)} \quad \text{[mi]}, \tag{5.26}$$

Since negative-sequence current does not exist during a three-phase balanced fault, this method cannot be used to locate a three-phase fault.

5.1.5 Girgis et al. Method

The method proposed by Girgis et al. [91] also accounts for system load when estimating the distance to fault. In addition, it allows the user to estimate the fault resistance.

It manipulates (5.6) by adding and subtracting the term $mz_1 I_{S_0}^f$ to yield:

$$V_{S_a}^f = mz_1\left(I_{S_1}^f + I_{S_2}^f + I_{S_0}^f\right) + mz_0 I_{S_0}^f - mz_1 I_{S_0}^f + I_{F_a}^f R_f \quad [\text{V}]. \tag{5.27}$$

On rearranging the terms, we get:

$$\begin{aligned} V_{S_a}^f &= mz_1 I_{S_a}^f + mI_{S_0}^f\left(z_0 - z_1\right) + I_{F_a}^f R_f \\ &= mz_1\left(I_{S_a}^f + kI_{S_0}^f\right) + I_{F_a}^f R_f \quad [\text{V}], \end{aligned} \tag{5.28}$$

where k, the zero-sequence compensation factor, is defined in Table 5.1. Dividing throughout by $\left(I_{S_a}^f + kI_{S_0}^f\right)$, we can calculate the apparent impedance measured by the relay during the A-G fault, Z_{app}:

$$Z_{app} = \frac{V_{S_a}^f}{I_{S_a}^f + kI_{S_0}^f} = mz_1 + R_f\left(\frac{I_{F_a}^f}{I_{S_a}^f + kI_{S_0}^f}\right) \quad [\Omega]. \tag{5.29}$$

Generalizing the above equation so that it is applicable to all faults, we get:

$$Z_{app} = \frac{V_S}{I_S} = mz_1 + R_f\left(\frac{I_F}{I_S}\right) \quad [\Omega], \tag{5.30}$$

where V_S and I_S are defined in Table 5.1. In the above equation, we cannot directly measure I_F as it would be quite a coincidence to have a meter or relay right at the fault point. We cannot substitute I_F with I_S as I_S is a summation of load current and I_F.

The workaround proposed by Girgis et al. combines the principles of the Takagi and the modified Takagi methods. It substitutes I_F with three times the zero-sequence current measured at the substation during a single line-to-ground fault. This is a good approximation of I_F as zero-sequence current exists only during a ground fault and not during load. For other fault types, the prefault current is simply subtracted out from the fault current measured at the substation to find the current at the fault point. The substitution is shown below, where I_F is replaced with I_{comp}, which is defined in Table 5.2 for all fault types.

$$Z_{app} = \frac{V_S}{I_S} = mz_1 + R_f\left(\frac{I_{comp}}{I_S}\right) \quad [\Omega]. \tag{5.31}$$

You can see that although we eliminated I_F from the impedance to fault equation, we still have one equation but two unknowns, namely, m and R_f. To level the playing field and get two equations, we write every complex quantity in (5.31) with their corresponding real and imaginary parts as shown below.

$$\begin{aligned} R_{app} + jX_{app} &= m\left(r_1 + jx_1\right) + R_f\left(\frac{I_d + jI_q}{I_{s1} + jI_{s2}}\right) \\ &= mr_1 + R_f L + j\left(mx_1 + R_f M\right) \quad [\Omega], \end{aligned} \tag{5.32}$$

where

$$R_{app} = \text{real}\left(\frac{V_S}{I_S}\right) \quad [\Omega],$$

$$X_{app} = \text{imag}\left(\frac{V_S}{I_S}\right) \quad [\Omega],$$

Table 5.2 Definition of I_{comp} for All Fault Types

Fault Type	I_{comp}
A-G	$3I_{S_0}^f$
B-G	$3I_{S_0}^f$
C-G	$3I_{S_0}^f$
AB or AB-G	$\left(I_{S_a}^f - I_{S_a}^p\right) - \left(I_{S_b}^f - I_{S_b}^p\right)$
BC or BC-G	$\left(I_{S_b}^f - I_{S_b}^p\right) - \left(I_{S_c}^f - I_{S_c}^p\right)$
CA or CA-G	$\left(I_{S_c}^f - I_{S_c}^p\right) - \left(I_{S_a}^f - I_{S_a}^p\right)$
ABC	same as line-to-line fault or double line-to-ground fault

$$L = \frac{I_d I_{s1} + I_q I_{s2}}{I_{s1}^2 + I_{s2}^2},$$

$$M = \frac{I_q I_{s1} - I_d I_{s2}}{I_{s1}^2 + I_{s2}^2}.$$

Equating only the real components gives us one equation:

$$R_{app} = mr_1 + R_f L$$

$$\therefore R_f = \frac{R_{app} - mr_1}{L} \quad [\Omega].$$

(5.33)

Equating only the imaginary components gives us a second equation, shown below:

$$X_{app} = mx_1 + R_f M \quad [\Omega].$$

(5.34)

Substituting (5.33) in (5.34), we solve for m:

$$m = \frac{X_{app}L - R_{app}M}{x_1 L - r_1 M} \quad [\text{mi}].$$

(5.35)

In summary, when implementing the Girgis method, first calculate fault location using (5.35). Then use (5.33) to calculate fault resistance.

5.1.6 Santoso et al. Method

The method proposed by Santoso et al. [92] solves for the minimum distance to fault. In other words, the estimate from this method can be treated as the lower bound of the distance estimate. Field personnel will find the fault either at this location or further down the line. The method writes the distance to fault equation given by (5.31) as a function of fault resistance, as shown below.

$$m = \frac{Z_{app} - R_f \times p_1}{z_1} \quad [\text{mi}],$$

(5.36)

where p_1 is defined below:

$$p_1 = \frac{I_{comp}}{I_S} = u_1 + jv_1. \tag{5.37}$$

Variables I_{comp} and I_S in the above equation are defined in Table 5.2 and Table 5.1, respectively. Writing the distance to fault equation with real and imaginary parts we get:

$$
\begin{aligned}
m &= \frac{R_{app} + jX_{app} - R_f\left(u_1 + jv_1\right)}{r_1 + jx_1} \\
&= \frac{\left(R_{app} - u_1R_f\right) + j\left(X_{app} - v_1R_f\right)}{r_1 + jx_1} \quad \text{[mi]}.
\end{aligned} \tag{5.38}
$$

Multiplying the numerator and denominator with $\left(r_1 - jx_1\right)$ and grouping all real and all imaginary parts together, we can express the above equation in its simplest form as

$$m = A + jB, \tag{5.39}$$

where A and B are defined below.

$$A = \frac{r_1\left(R_{app} - u_1R_f\right) + x_1\left(X_{app} - v_1R_f\right)}{r_1^2 + x_1^2},$$

$$B = \frac{r_1\left(X_{app} - v_1R_f\right) - x_1\left(R_{app} - u_1R_f\right)}{r_1^2 + x_1^2}.$$

Taking the magnitude, we get:

$$|m| = \sqrt{A^2 + B^2} = \sqrt{\frac{aR_f^2 + bR_f + c}{r_1^2 + x_1^2}}, \tag{5.40}$$

where

$$a = u_1^2 + v_1^2,$$
$$b = (-2) \times \left(u_1R_{app} + v_1X_{app}\right),$$
$$c = R_{app}^2 + X_{app}^2.$$

The numerator of (5.40) is a quadratic function that always possesses a minimum value since $a > 0$. The minimum value occurs when R_f equals the value shown below.

$$R_f = \frac{-b}{2a} = \frac{u_1R_{app} + v_1X_{app}}{u_1^2 + v_1^2} \quad [\Omega] \tag{5.41}$$

Substituting (5.41) in (5.40), we can solve for m:

$$|m| = \frac{u_1X_{app} - v_1R_{app}}{|z_1||p_1|} \quad \text{[mi]}. \tag{5.42}$$

5.1.7 Novosel et al. Method

The method proposed by Novosel et al. [93] assumes all loads to be of a constant impedance nature and lumps them at the end of the feeder. Dividing the positive-sequence voltage at the substation, $V_{S_1}^p$, with the positive-sequence current at the substation, $I_{S_1}^p$, during normal

Figure 5.11 Positive-sequence pure fault network during a fault.

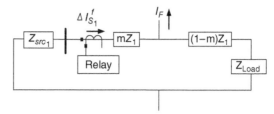

load conditions gives us an impedance that equals the sum of the line impedance and the load impedance. The load impedance can then be calculated as:

$$Z_{Load} = \frac{V_{S_1}^p}{I_{S_1}^p} - Z_1 \quad [\Omega], \tag{5.43}$$

where Z_1 is the total line impedance in ohms. We learned about pure fault networks in Chapter 3. Figure 5.11 shows the pure fault positive-sequence network during a fault. In the figure, m represents the per unit distance to the fault. Since the load impedance is now known, we can calculate I_F using the current division rule:

$$I_F = \frac{Z_{Load} + Z_1 + Z_{src_1}}{(1-m) \times Z_1 + Z_{Load}} \times \Delta I_S \quad [A], \tag{5.44}$$

where ΔI_S equals $\Delta I_{S_1}^f$. Substituting I_F in (5.30), we get:

$$\begin{aligned}
\frac{V_S}{I_S} &= mz_1 + R_f \left(\frac{I_F}{I_S} \right) \\
&= mz_1 + R_f \times \frac{\Delta I_S}{I_S} \times \left(\frac{Z_{Load} + Z_1 + Z_{src_1}}{(1-m) \times Z_1 + Z_{Load}} \right) \quad [\Omega].
\end{aligned} \tag{5.45}$$

After a fairly lengthy manipulation, we end up with the simplified expression below:

$$m^2 - k_1 m + k_2 - k_3 R_f = 0, \tag{5.46}$$

where the constants are defined as

$$k_1 = a + jb = 1 + \frac{Z_{Load}}{Z_1} + \left(\frac{V_S}{Z_1 \times I_S} \right),$$

$$k_2 = c + jd = \frac{V_S}{Z_1 \times I_S} \left(1 + \frac{Z_{Load}}{Z_1} \right),$$

$$k_3 = e + jf = \frac{\Delta I_S}{Z_1 \times I_S} \left(1 + \frac{Z_{Load} + Z_{src_1}}{Z_1} \right).$$

Decomposing (5.46) into its real and imarginary parts, we further simplify to get:

$$\left(m^2 - am + c - eR_f \right) + j \left(d - bm - fR_f \right) = 0. \tag{5.47}$$

Equating the imaginary part to 0, we derive an expression for R_f:

$$d - bm - fR_f = 0,$$

$$R_f = \frac{d - bm}{f} \quad [\Omega]. \tag{5.48}$$

Equating the real part to 0 and substituting the expression for R_f, we get the following quadratic equation:

$$m^2 - am + c - eR_f = 0,$$

$$m^2 - am + c - e\left(\frac{d - bm}{f}\right) = 0, \tag{5.49}$$

$$m^2 - m\left(a - \frac{eb}{f}\right) + \left(c - \frac{ed}{f}\right) = 0.$$

The distance to fault can be solved as:

$$m = \frac{\left(a - \frac{eb}{f}\right) \pm \sqrt{\left(a - \frac{eb}{f}\right)^2 - 4\left(c - \frac{ed}{f}\right)}}{2} \times l \quad [\text{mi}]. \tag{5.50}$$

The value of m between 0 and the line length should be chosen as the location estimate. If the local source impedance Z_{src_1} is not known, it can be estimated from the negative-sequence network using the equation below:

$$Z_{src_1} = -\frac{V_{S_2}^f}{I_{S_2}^f} \quad [\Omega]. \tag{5.51}$$

Here $V_{S_2}^f$ and $I_{S_2}^f$ are the negative-sequence voltage and current recorded by a relay at the substation during an unbalanced fault.

5.2 Challenges with Distribution Fault Location

5.2.1 Load

The simple reactance and loop reactance methods neglect load when computing the distance to fault. When the fault is bolted, meaning it has zero resistance, this assumption does not affect fault location accuracy even if the distribution feeder is significantly loaded. However, if the fault has some resistance, then this assumption will lead to a reactance error as described in Chapter 3. The error is worst for faults at the very end of the feeder, where the fault current is extremely low and difficult to separate from load current. Use the Takagi, modified Takagi, Girgis et al., Santoso et al., or the Novosel et al. method to locate faults on feeders that are highly loaded as those methods account for system load.

5.2.2 Non-Homogeneous Lines

Impedance-based fault location algorithms assume that the distribution feeder is homogeneous, i.e., it is made up of the same type of conductor and it has the same pole configuration throughout the length of the feeder. In reality, a distribution feeder can be made up of multiple conductor types. The pole configuration can also change during the feeder run. For example, Figure 5.12 shows a distribution pole in which the conductors are arranged horizontally while Figure 5.13 shows the phase conductors arranged vertically. As a result, z_1 and z_0 are different for every line section. The question is, what value of z_1 and z_0 do you use for fault location?

Figure 5.12 Distribution pole with all three phase conductors arranged horizontally on a crossarm.

One reference [72] suggests using the line impedance of the most commonly used conductor. For example, consider an eight-mile-long distribution feeder. Suppose that the first five miles of the feeder are made up of 1/0 ACSR phase and 1/0 ACSR neutral conductors. The positive- and zero-sequence line impedance per mile of that line section are:

$$z_1 = 1.1201 + j0.8333 \quad [\Omega/\text{mi}],$$
$$z_0 = 1.7498 + j2.3717 \quad [\Omega/\text{mi}].$$

The remaining three miles of the distribution feeder are made up of #2 6/1 ACSR phase and #2 6/1 ACSR neutral conductor. The positive- and zero-sequence line impedance per mile of that line section are:

$$z_1 = 1.3681 + j0.8412 \quad [\Omega/\text{mi}],$$
$$z_0 = 2.0380 + j2.4721 \quad [\Omega/\text{mi}].$$

Use the 1/0 ACSR line impedance parameters when calculating fault location.

Figure 5.13 Distribution pole with all three phase conductors arranged vertically.

Another reference [51] suggests using a nomograph to correct the error due to a non-homogeneous distribution feeder. This graph takes the estimate from an impedance-based fault location algorithm and converts it to a more realistic location by accounting for feeder non-homogeneity. To build a nomograph, model the distribution feeder in CAPE, PSCAD, ASPEN OneLiner, or any other short-circuit program. Then simulate a fault at a known location and record the voltage and current output from the short-circuit program. Estimate fault location using the recorded voltage and current values. Plot the actual location versus the estimated location as shown in Figure 5.14. Repeat the above steps by simulating more faults along the length of the feeder.

Another reference [94] suggests using the loop reactance method to calculate the reactance to fault instead of the distance to fault to account for a non-homogeneous feeder. As an example, the reactance to fault (X_{fault}) is the numerator of (5.10) for a single line-to-ground fault. Compare the reactance to fault with the reactance of the first line section. If the reactance to fault is greater, then the fault is beyond the first line section. Repeat this process for the other line sections until you find a line section in which X_{fault} is greater than the

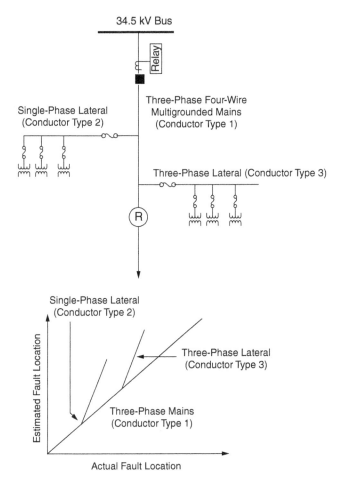

Figure 5.14 Nomograph shown for an example feeder.

reactance from the measurement point to the beginning of the line section but less than the reactance from the measurement point to the end of the line section. Once the faulted section is identified, use the line impedance parameters of that particular line section to calculate fault location as shown below:

$$m = L - \frac{X - X_{fault}}{x},$$ (5.52)

where L is the distance between the measurement point and the end of the identified faulted line section, X is the loop reactance between the measurement point and the end of the identified faulted line section, and x is the reactance of the faulted section in ohms per mile.

5.2.3 Inaccurate Earth Resistivity

Impedance-based fault location algorithms require knowledge of the zero-sequence line impedance when estimating the distance to a single line-to-ground fault. To calculate the zero-sequence line impedance, we need to know the earth resistivity. Unfortunately, the exact value of the earth resistivity is never known accurately as it varies with soil type, temperature, season, and moisture content in the soil. As a result, an inaccurate value of the zero-sequence line impedance will affect the accuracy of the impedance-based fault location algorithms when locating single line-to-ground faults. This is discussed in details in Chapter 3.

5.2.4 Multiple Laterals

The distribution network is extensive as multiple single- and three-phase laterals are tapped from the main feeder to serve loads as described in the beginning of this chapter. These laterals pose a major challenge to fault location as they give rise to multiple possibilities of where the fault could be located. As an example, suppose that the distribution feeder shown in Figure 5.15 experienced a fault. Using the data recorded by the relay at the substation, the distance to fault is estimated to be a certain number of miles from the substation. When this distance is mapped on the distribution feeder, you can see that it gives rise to multiple possible locations. The fault could be on the single-phase lateral, on the three-phase lateral, or on the main feeder as shown in Figure 5.15. We do not encounter this problem in transmission lines as each line has a protective relay or some sort of a recording device that captures data during a fault. However, in distribution networks, a relay or a recording device is present only at the substation. Sometimes a recloser, installed mid-line to improve selectivity, can help narrow down fault location.

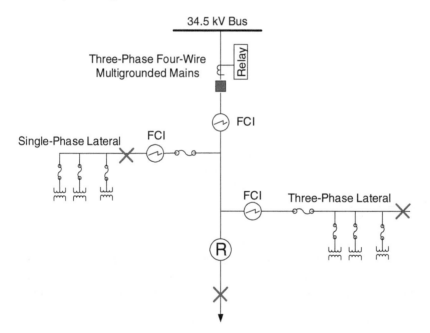

Figure 5.15 Using data from the substation relay to estimate fault location can give rise to multiple possibilities. One option is to use FCIs to narrow down the location.

Figure 5.16   FCI devices hanging on the line.

Using faulted circuit indicators (FCI in Figure 5.15) can also help narrow down the location of the fault. These indicators hang on the overhead line (see Figure 5.16) and provide a visual indication of a fault. Linemen can follow these FCIs and identify the faulted section. Figure 5.17 shows an example FCI device that not only provides a visual indicator of a fault but also indicates whether the fault was permanent or temporary by the color of the flashing LED. In addition, the device transmits the fault status to SCADA or an outage management system. This provides line crew with crucial information on which FCI devices saw the fault and which section is faulted without actually patrolling the line. This reduces outage time and saves money.

Knowing the phase that has faulted also helps. Suppose that the feeder experienced a B-G fault. The single-phase lateral, on the other hand, is A-phase. This rules out the possibility that the fault is on the single-phase lateral. Knowing how long it took to clear the fault and comparing that information with a database that contains all protective device timing information can also help pinpoint the fault location.

5.2.5   Best Data for Fault Location: Feeder or Substation Relays

Suppose that a fault occurs downstream from the recloser in Figure 5.15. Both the relay at the substation and the recloser saw the same fault and recorded voltage and current during the fault. Data from which device would be the best to use for fault location? Using data

Figure 5.17 An example of a faulted circuit indicator. The bright LEDs light up during a fault that can be seen from 50 m away during the day and from 100 m at night. (*Photo: Courtesy of Schweitzer Engineering Laboratories.*)

from the device closest to the fault will yield the highest accuracy as the impact of some of the error sources such as load and non-homogeneous lines can be minimized.

5.2.6 Distributed Generation

Distributed generation consists of generators directly connected to the distribution grid. Many of these generators harvest energy from biogas, sun, wind, and other renewable sources of energy. As a result, there has been a lot of investment in distributed generation over the last couple of years as governments across the world are pushing for a future with clean energy. With distributed generation, however, distribution feeders are no longer radial in nature. This negatively affects the accuracy of impedance-based fault location algorithms as they have been developed for a conventional distribution feeder that is radial all the way from the substation to the load. When a fault occurs downstream from the DG unit, short-circuit current to a fault comes from two sources, the utility substation and distributed generators. This modifies the apparent impedance seen from the substation.

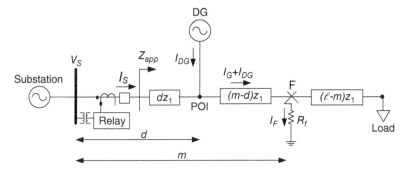

Figure 5.18 Distribution feeder with a fault located downstream from a DG unit.

To illustrate the concept described above, consider the distribution feeder shown in Figure 5.18. The feeder has a positive-sequence line impedance of z_1 ohms per mile. The DG is interconnected to the grid (referred to as the point of interconnection, or POI) at a distance of d miles from the substation. When a three-phase fault with a fault resistance R_f occurs at a distance of m miles from the substation, the substation and DG operate in parallel to feed the fault. Voltage drop from the substation can be written as:

$$V_S = dz_1 I_S + (m - d)z_1 \left(I_S + I_{DG}\right) + R_f I_F, \tag{5.53}$$

where V_S and I_S are the voltage and current phasors recorded at the substation during the fault, I_{DG} is the fault current from the DG unit, and I_F is the current at the fault point. Dividing throughout by I_S and simplifying, the apparent impedance (Z_{app}) seen from the substation is

$$Z_{app} = \frac{V_S}{I_S} = mz_1 + (m - d)z_1 \frac{I_{DG}}{I_S} + R_f \left(\frac{I_F}{I_S}\right). \tag{5.54}$$

When the fault is located downstream from the DG unit, the apparent impedance seen from the substation is proportional to the impedance to the fault (mz_1) as well as two additional terms: $(m - d)z_1 \left(I_{DG}/I_S\right)$ and $R_f \left(I_F/I_S\right)$. Since one-ended fault location methods make use of only V_S and I_S at the substation, neglecting I_{DG} in (5.54) will certainly compromise the accuracy of the location estimates.

For a bolted fault, the term $(m - d)z_1 \left(I_{DG}/I_S\right)$ increases Z_{app} as shown in Figure 5.19 (a). As a result, impedance-based fault location algorithms will overestimate the distance to the location of the fault. When the fault has a significant R_f, impedance-based fault location algorithms are affected by an additional reactance error. Since short-circuit current at the fault point (I_F) comes from the utility substation (I_S) and the distributed generator (I_{DG}), the phase angles of I_F and I_S are not equal to each other. When I_F leads I_S, the term $R_f \left(I_F/I_S\right)$ is inductive and together with $(m - d)z_1 \left(I_{DG}/I_S\right)$ increases the apparent impedance to the fault as shown in Figure 5.19 (b). This is the worst-case scenario and fault location algorithms will significantly overestimate the distance to the fault location. When I_F lags I_S, on the other hand, $R_f \left(I_F/I_S\right)$ is capacitive and will attempt to cancel out the effect of $(m - d)z_1 \left(I_{DG}/I_S\right)$ as illustrated in Figure 5.19 (c).

So far, we have learned about the negative impact of DG when the fault is located downstream from it. What if a fault occurs upstream from the DG unit, between the utility substation and the distributed generator? In that case, the scenario is similar to a fault occurring

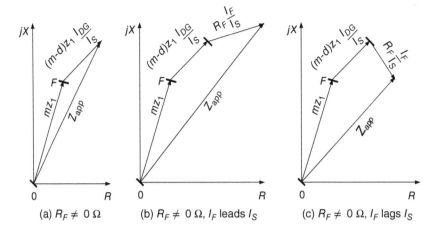

Figure 5.19 Apparent impedance Z_{app} from the substation.

on a two-terminal transmission line. The performance of the impedance-based methods will be much improved as they will be affected only by the reactance error.

Clearly faults occurring downstream from the DG unit have the most adverse impact on impedance-based fault location algorithms. Reference [95] evaluates some of the factors that either help or degrade the performance of impedance-based fault location algorithms when locating faults downstream from a DG unit. The first factor is the type of DG technology. The magnitude and duration of the fault current contributed by distributed generators (I_{DG}) depends on the electrical converter that DGs use to interconnect to the grid. As established by IEEE Standard 1547 [96], there are three types of electrical converters: synchronous machines, induction machines, and inverter-based DGs. Synchronous machines, used by diesel generators, gas turbines and hydro generators, provide a sustained fault current and have the worst impact on impedance-based fault locating algorithms [97]. Fault current is 5 to 10 times the rated current during the sub-transient period and then decays to 2 to 4 times the rated current [98]. Induction generators, used in fixed-speed and wide-slip wind turbines, can also contribute to a fault so long as the residual voltage on the healthy phase can establish a rotating magnetic field. Although the initial magnitude of the fault current is 5 to 10 times the rated current, the current decays at a rate that depends on the fault type [99]. Therefore, the error in fault location will also depend on the fault type. Inverter-based DGs such as fuel cells, photovoltaic generators, double-fed induction generators (DFIGs), and permanent magnet wind turbines contribute fault current for less than half a cycle. In the worst case, if inverter-based DGs continue to feed a fault, the fault current is only 1 to 2 times the rated current. As a result, the corresponding error in fault location estimates is lower than other DG technologies [100].

The second factor that was evaluated was the type of transformer used to interface DGs to the grid. If the transformer configuration is a delta on the grid side, then DGs will not contribute any fault current during a single line-to-ground fault. Recall that single line-to-ground faults are the most common among all other types of fault.

The third factor was the size of the DG unit. In (5.54), the magnitude of I_{DG} in the terms $(m - d)Z_{L1}(I_{DG}/I_G)$ and $R_f(I_F/I_G)$ depends on the MVA capacity of the installed distributed generators. Single DG units having a small MVA capacity, they do not contribute a significant fault current. However, when the MVA capacity is increased by aggregating a number of small or a few large DG units, the equivalent generator impedance decreases [98]. As a result, the total fault current from DGs increase and can significantly offset the accuracy of fault location algorithms.

The fourth factor was the location of the fault from the DG unit. Location of the fault m affects the magnitude of this term in two ways: factors $(m - d)$ and the current ratio I_{DG}/I_G. When a fault occurs at the POI, the fault is very close to the distributed generator. Current I_{DG} and, therefore, I_{DG}/I_G are maximum. However, because m is equal to d, the term $(m - d)Z_{L1}(I_{DG}/I_G)$ becomes zero, and DG will not affect the accuracy of location estimates. When the same fault is simulated further downstream from DGs, $(m - d)$ increases. The current ratio I_{DG}/I_G, on the other hand, decreases since impedance to the fault increases. However, the increase in $(m - d)$ is the dominating factor and increases the error in fault location.

The final factor that was evaluated was the effect of tapped loads. Impedance-based fault location algorithms assume that the load served by a distribution feeder is lumped at the end of the feeder. In practice, loads are tapped along the entire length of the feeder as shown in Figure 5.20. When a bolted fault ($R_f = 0 \, \Omega$) occurs at point F shown in Figure 5.20, the actual impedance between the substation and fault point F (Z_{SF}) is

$$Z_{SF} = Z_{L1} + Z_{L2}, \tag{5.55}$$

where Z_{L1} is the positive-sequence impedance of the line segment between the substation and the tap point, and Z_{L2} is the positive-sequence line impedance between the tap point and the fault. However, due to tapped loads, the apparent impedance measured from the substation comes out to be

$$Z_{app} = \frac{V_S}{I_S} = Z_{L1} + \frac{Z_{L2} \times Z_{Load}}{Z_{L2} + Z_{Load}} < Z_{SF}. \tag{5.56}$$

Since Z_{Load} has a much higher impedance than Z_{L2}, the resultant value of $(Z_{L2} \times Z_{Load})/(Z_{L2} + Z_{Load})$ in (5.56) is closer to but slightly smaller than Z_{L2} [101]. In other words, tapped loads act as a negative impedance, and Z_{app} is smaller than the actual impedance to the fault, Z_{SF}. In contrast, fault current contribution from DGs (I_{DG})

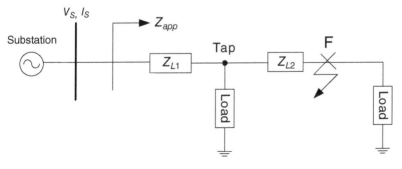

Figure 5.20 Load tapped along the entire length of the distribution feeder.

increase Z_{app} seen from the substation. Therefore, the two error factors, tapped load and I_{DG}, tend to cancel each other out and can help improve the accuracy of impedance-based fault location algorithms.

5.2.7 High Impedance Faults

A high impedance fault is one during which the fault has an extremely high impedance and during which fault current magnitude is small [102]. An example of a high impedance fault is when a phase conductor falls on a poorly conducting surface such as concrete or dry grass. Reference [103] staged a high impedance fault by lowering an energized conductor on different test surfaces at a distance of 12.7 miles from the substation. On contact with concrete, the fault current at the substation was measured to be 10 A. When the energized conductor made contact with a tree, the fault current at the substation was measured to be around 16 A. High impedance faults also involve a lot of arcing, which introduce harmonics in the current measurement. The small fault current magnitude together with harmonics make the task of locating faults via impedance-based methods extremely challenging [104]. Dirty insulators and trees touching energized conductors are also potential causes of high impedance faults.

5.2.8 CT Saturation

Fault location algorithms require the input of accurate current phasors during the fault to estimate fault location. For this purpose, they rely on current transformers. These transformers measure the primary current and scale it down to a secondary current that is safe for a protective relay or any other recording device. In an ideal world, the scaled secondary current is an exact replica of the primary current. However, if the fault has a significant DC offset, the magnetic core of the current transformer may saturate, and the secondary current does not replicate the primary. This is known as asymmetrical saturation [74]. Using the distorted output from a saturated CT will lead to an error in fault location. As the DC offset decays, the CT starts to come out of saturation. The best way to handle asymmetrical saturation is to wait a couple of cycles till the CT comes out of saturation and then perform fault location. Another type of CT saturation is known as symmetrical saturation. In this case, the CT is not sized correctly and cannot handle a significant amount of primary symmetrical fault current magnitude. The CT will remain saturated during the duration of the fault, leading to inaccurate fault location.

5.2.9 Grounding

The type of grounding in a distribution network plays an important role in locating single line-to-ground faults. Distribution networks may be solidly grounded, ungrounded, resistance grounded, or grounded through a Peterson coil. Most distribution networks in North America are three-phase four-wire systems in which the neutral is grounded at multiple points as shown in Figure 5.21. Impedance-based fault location algorithms will have the highest accuracy in these systems. In ungrounded systems, grounding transformers are installed, which act as a source of ground current during a ground fault.

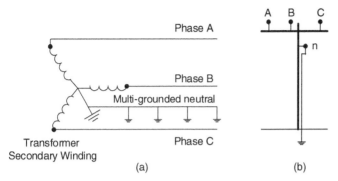

Figure 5.21 (a) Three-phase four-wire multi-grounded neutral system. (b) Utility pole configuration in which the neutral conductor is grounded at the base of the pole [89].

The ground current magnitude though is much reduced due to fewer number of ground current sources [36]. Resistance grounded networks and those that are grounded via Peterson coils will also also have very low magnitude of ground fault current. Applying impedance-based fault location algorithms to these three types of networks will reduce the accuracy of these methods.

5.2.10 Short Duration Faults

Impedance-based fault location algorithms use voltage and current phasors during a fault to calculate fault location. A filter such as the fast Fourier transform is used to extract the required phasors during the fault. This filter requires that the fault last for at least one cycle so that the phasors can be calculated accurately. If the fault is a blip and self-clears within a cycle, impedance-based fault location algorithms are unable to calculate fault location. An example is shown in Figure 5.22. A squirrel made contact with a B-phase line switch that caused a half-cycle blip. Figure 5.23 shows another example where a tree momentarily touched an energized conductor. The fault lasts a little longer, about a cycle. But notice that the positive and negative half cycle of the fault current have different peak magnitudes. This indicates that the fault resistance was varying during this short duration fault. Impedance-based fault location algorithms would be unsuccessful in determining location of such short duration faults.

5.2.11 Missing Voltage

Distribution feeders are normally protected by overcurrent relays. Data recorded by these devices during a fault are usually used for fault location purposes. In these distribution relays, voltage measurements are optional. Furthermore, voltage transformers are protected by fuses. If one of the fuses blow, the relay cannot measure the corresponding phase voltage. All the fault location algorithms that were discussed in this chapter require both voltage and current during a fault to estimate the fault location. As a result, if voltage is missing or not available, we cannot estimate the fault location. In Chapter 6, we discuss how we may overcome this problem by using current measurements alone to estimate the distance to a fault.

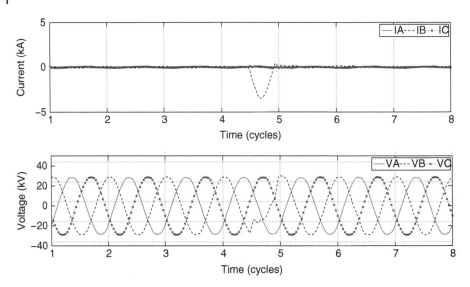

Figure 5.22 Squirrel makes momentary contact with the B-phase line switch.

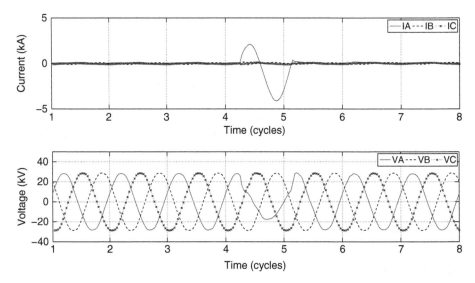

Figure 5.23 Tree makes momentary contact with the A-phase conductor.

5.3 Exercise Problems

■ Exercise 5.1

A 34.5 kV distribution feeder was struck by lightning and experienced a fault. The high fault current triggered a recording device at the substation to record a 16-cycle long event report with a resolution of 32 samples per cycle as shown in Figure 5.24. This event report is a

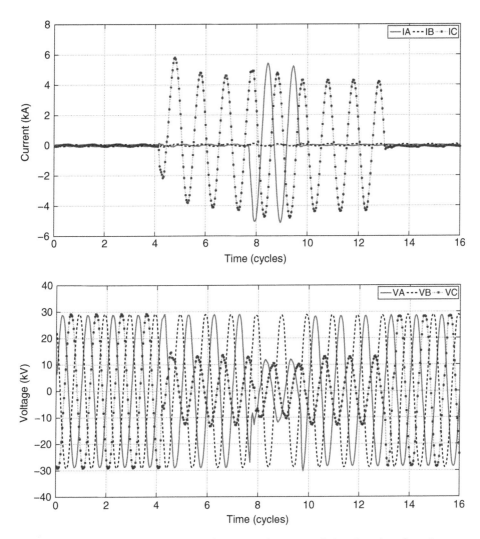

Figure 5.24 Exercise 5.1: Current and voltage waveforms recorded at the substation when lightning struck a distribution feeder and caused a C-G fault. During the middle of the fault, the A-phase also became involved in the fault.

stored record of the three-phase line currents and line-to-neutral voltages that the recording device measured at the time of the lightning strike and is what helped piece together the sequence of events. It can be observed that up until the fourth cycle, currents and voltages looked healthy. However, when lightning struck the line, the resulting voltage surge exceeded the insulation flashover strength of phase C and created a C-phase-to-ground fault. The A-phase also became involved in the fault at around the eighth cycle, momentarily, for about two cycles, creating a CA-G fault. We know that the ground is involved as the two faulted phases are not exactly out of phase with each other. The C-phase-to-ground fault was eventually cleared by the operation of a protective device closest to the fault. The utility would like to locate the flashover point. Which cycle would be best for calculating fault location?

Solution:

It is important to be strategic when choosing the one-cycle window for fault location. Choosing a cycle toward the middle of the fault is ideal and will yield the best fault location results as the voltages and currents are expected to have stabilized by then. In this event, the fault starts just a little after the fourth cycle and is cleared by the thirteenth cycle. Notice that during the first couple of cycles, the fault has a significant DC offset and will not be an ideal candidate for fault location. We want to avoid the tail end of the fault as well because of the filter effect described in Chapter 4. Between cycle 7.5 and cycle 10, both A and C phases become involved in the fault. Data during this CA-G fault could have been used for fault location except we don't know whether the C-G fault evolved into a CA-G fault at the same location or whether the distribution feeder was experiencing two simultaneous faults, a C-G fault at one location and another A-G fault at another location. Remember that the root cause of this fault was lightning and that lightning can cause flashover at multiple locations. Impedance-based fault location algorithms have been developed based on the assumption that there is only one fault on the system. Trying to apply these algorithms to simultaneous faults can skew fault location accuracy. Therefore, all facts considered, the best window for fault location would be between the tenth and the twelfth cycle.

■ Exercise 5.2

The previous example concluded that the best window for fault location would be between the tenth and the twelfth cycle. A fast Fourier transform filter was used to extract the voltage and current phasors at the eleventh cycle. They are given below.

$$
I^f_{S_{abc}} [A] \qquad V^f_{S_{abc}} [kV]
$$

$$
\begin{bmatrix} 40.10\angle -91.92° \\ 72.90\angle 56.39° \\ 4339.70\angle 75.46° \end{bmatrix} \quad \begin{bmatrix} 28.62\angle -88.95° \\ 29.23\angle 31.20° \\ 12.32\angle 138.81° \end{bmatrix}.
$$

Voltage and current phasors at the second cycle (prefault) are given below.

$$
I^p_{S_{abc}} [A] \qquad V^p_{S_{abc}} [kV]
$$

$$
\begin{bmatrix} 61.20\angle -49.22° \\ 66.50\angle 65.89° \\ 65.60\angle 174.98° \end{bmatrix} \quad \begin{bmatrix} 29.19\angle -88.43° \\ 29.55\angle 31.51° \\ 29.53\angle 151.50° \end{bmatrix}.
$$

The prefault voltage phasors are drawn in Figure 5.25. Note that the phasors are not drawn to scale. Looking at the phasor diagram, one can tell that the utility in question has ACB phase rotation. The distribution feeder is non-homogeneous with the most commonly occurring conductor type having a positive- and zero-sequence impedance of $z_1 = 0.1308 + j0.5546$ Ω/mile and $z_0 = 0.4029 + j1.8619$ Ω/mile, respectively. Given the above data, determine the fault location using the following methods:

Figure 5.25 Exercise 5.2: prefault phasors

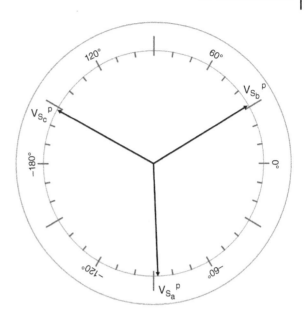

(a) Loop reactance method
(b) Simple reactance method
(c) Takagi method
(d) Modified takagi method
(e) Girgis et al. method
(f) Santoso et al. method

Solution:

All the fault location methods require knowledge of the zero-, the negative-sequence current phasor, or both. Let us calculate those required phasors first before solving for the fault location. We use the equation below to calculate the sequence currents, keeping in mind that the system has ACB phase rotation and that the C-phase must be used as the reference for a C-G fault.

$$
\begin{bmatrix} I^f_{S_0} \\ I^f_{S_1} \\ I^f_{S_2} \end{bmatrix} = \frac{1}{3} \begin{bmatrix} 1 & 1 & 1 \\ 1 & a & a^2 \\ 1 & a^2 & a \end{bmatrix} \times \begin{bmatrix} I^f_{S_c} \\ I^f_{S_b} \\ I^f_{S_a} \end{bmatrix}
$$

$$
= \frac{1}{3} \begin{bmatrix} 1 & 1 & 1 \\ 1 & a & a^2 \\ 1 & a^2 & a \end{bmatrix} \times \begin{bmatrix} 4339.70\angle 75.46° \\ 72.90\angle 56.39° \\ 40.10\angle -91.92° \end{bmatrix}
$$

$$
= \begin{bmatrix} 1456.50\angle 75.03° \\ 1446.40\angle 76.91° \\ 1437.50\angle 74.43° \end{bmatrix} \quad [A].
$$

We will also need to calculate the zero-sequence compensation factor, k:

$$k = \frac{z_0 - z_1}{z_1} = \frac{(0.4029 + j1.8619) - (0.1308 + j0.5546)}{0.1308 + j0.5546} = 2.3432\angle1.52°.$$

Finally, we will also need to calculate current I_S. Per Table 5.1, I_S for a C-G fault can be calculated as

$$I_S = I_{S_c}^f + kI_{S_0}^f$$
$$= 4339.70\angle75.46° + 2.3432\angle1.52° \times 1456.50\angle75.03°$$
$$= 7752.23\angle75.94° \quad [A].$$

Now, we are ready to apply the impedance-based fault location algorithms.

(a) The loop reactance method uses (5.10) to calculate fault location estimate as:

$$m = \frac{\text{imag}\left(\dfrac{V_{phase}}{I_0}\right)}{\text{imag}\left(2z_1 + z_0\right)}$$

$$= \frac{\text{imag}\left(\dfrac{12.32\angle138.81° \times 1000}{1456.50\angle75.03°}\right)}{\text{imag}\left(2 \times (0.1308 + j0.5546) + (0.4029 + j1.8619)\right)}$$

$$= 2.55 \quad [\text{mi}].$$

The estimated fault location of 2.55 miles is very close to the flashover point found by the utility at 2.67 miles from the substation. The lightly loaded condition of the distribution feeder at the time of the fault, load current being around 65 A, contributed to the excellent accuracy from this method.

(b) The simple reactance method uses (5.24) to calculate fault location. The calculation steps are shown below.

$$m = \frac{\text{imag}\left(\dfrac{V_S}{I_S}\right)}{\text{imag}\left(z_1\right)}$$

$$= \frac{\text{imag}\left(\dfrac{12.32\angle138.81° \times 1000}{7752.23\angle75.94°}\right)}{0.5546}$$

$$= 2.55 \quad [\text{mi}].$$

The simple reactance method performed equally well as the loop reactance method. Like the loop reactance method, this method also ignores load. However, because the distribution feeder was lightly loaded, this assumption did not significantly affect fault location accuracy.

(c) The Takagi method accounts for system load by subtracting out the prefault current as shown below:

$$\Delta I_S = I_{S_c}^f - I_{S_c}^p$$
$$= 4339.70\angle75.46° - 65.60\angle174.98°$$
$$= 4351.03\angle74.61° \quad [A].$$

Next, we use (5.25) to calculate fault location as

$$m = \frac{\text{imag}\left(V_S \times \Delta I_S^*\right)}{\text{imag}\left(z_1 \times I_S \times \Delta I_S^*\right)}$$

$$= \frac{\text{imag}\left(12.32\angle138.81° \times 1000 \times \left(4351.03\angle74.61°\right)^*\right)}{\text{imag}\left((0.1308 + j0.5546) \times 7752.23\angle75.94° \times \left(4351.03\angle74.61°\right)^*\right)}$$

$$= 2.57 \quad [\text{mi}].$$

Because the Takagi method accounts for load, there is a slight improvement in the accuracy of the fault location estimate.

(d) The modified Takagi method does not require prefault current information. Instead, it uses the negative-sequence current to account for load and calculate fault location as shown below:

$$m = \frac{\text{imag}\left(V_S \times I_{seq}^*\right)}{\text{imag}\left(z_1 \times I_S \times I_{seq}^*\right)}$$

$$= \frac{\text{imag}\left(12.32\angle138.81° \times 1000 \times \left(1437.50\angle74.43°\right)^*\right)}{\text{imag}\left((0.1308 + j0.5546) \times 7752.23\angle75.94° \times \left(1437.50\angle74.43°\right)^*\right)}$$

$$= 2.57 \quad [\text{mi}].$$

(e) To use the Girgis et al. method, let us first calculate the variables below.

$$R_{app} = \text{real}\left(\frac{V_S}{I_S}\right)$$

$$= \text{real}\left(\frac{12.32\angle138.81° \times 1000}{7752.23\angle75.94°}\right)$$

$$= 0.7247 \quad \Omega,$$

$$X_{app} = \text{imag}\left(\frac{V_S}{I_S}\right)$$

$$= \text{imag}\left(\frac{12.32\angle138.81° \times 1000}{7752.23\angle75.94°}\right)$$

$$= 1.4144 \quad \Omega,$$

$$I_d = \text{real}\left(I_{comp}\right) = \text{real}\left(3I_{S_0}^f\right) = 1128.70 \quad [\text{A}],$$

$$I_q = \text{imag}\left(I_{comp}\right) = \text{imag}\left(3I_{S_0}^f\right) = 4221.20 \quad [\text{A}],$$

$$I_{s1} = \text{real}\left(I_S\right) = \text{real}\left(7752.23\angle75.94°\right) = 1883.31 \quad [\text{A}],$$

$$I_{s2} = \text{imag}\left(I_S\right) = \text{imag}\left(7752.23\angle75.94°\right) = 7519.99 \quad [\text{A}],$$

$$L = \frac{I_d I_{s1} + I_q I_{s2}}{I_{s1}^2 + I_{s2}^2}$$

$$= \frac{1128.70 \times 1883.31 + 4221.20 \times 7519.99}{1883.31^2 + 7519.99^2} = 0.5636,$$

$$M = \frac{I_q I_{s1} - I_d I_{s2}}{I_{s1}^2 + I_{s2}^2}$$

$$= \frac{4221.20 \times 1883.31 - 1128.70 \times 7519.99}{1883.31^2 + 7519.99^2} = -0.00895.$$

The distance to fault can then be calculated as:

$$m = \frac{X_{app}L - R_{app}M}{x_1 L - r_1 M}$$

$$= \frac{1.4144 \times 0.5636 - 0.7247 \times (-0.00895)}{0.5546 \times 0.5636 - 0.1308 \times (-0.00895)}$$

$$= 2.56 \quad [\text{mi}].$$

The fault resistance is estimated to be 0.69 Ω as shown below:

$$R_f = \frac{R_{app} - mr_1}{L}$$

$$= \frac{0.7247 - 2.56 \times 0.1308}{0.5636}$$

$$= 0.69 \quad \Omega.$$

(f) The Santoso et al. method calculates the lower bound of the distance estimate. When using this method, the first step is to calculate variable p_1 using (5.37):

$$p_1 = \frac{I_{comp}}{I_S} = \frac{3I^f_{S_0}}{I^f_{S_c} + kI^f_{S_0}}$$

$$= \frac{3 \times 1456.50\angle 75.03°}{7752.23\angle 75.94°}$$

$$= 0.5636 - j0.0089,$$

$$\therefore u_1 = \text{real}\,(p_1) = 0.5636$$

$$\text{and} \quad v_1 = \text{imag}\,(p_1) = -0.0089.$$

The distance to fault can then be solved using (5.42):

$$|m| = \frac{u_1 X_{app} - v_1 R_{app}}{|z_1||p_1|}$$

$$= \frac{0.5636 \times 1.4144 + 0.0089 \times 0.7247}{0.5698 \times 0.5637}$$

$$= 2.50 \quad [\text{mi}].$$

In summary, all methods performed exceptionally well with estimates close to the actual location of the fault, which is 2.67 miles from the substation.

■ **Exercise 5.3**

A protective relay protecting a 12 kV distribution feeder opened the station breaker. The 2.65-mile-long distribution feeder was serving 328 customers, all of whom lost power. The utility is receiving calls from several upset customers wanting an early reprieve from the sweltering heat of May. Frantic, the utility wants to apply fault location algorithms and get a head start on where to start looking for the fault. The sooner they find the fault, the sooner they can restore power back to their customers. The utility was able to communicate with the relay at the substation and download the 11-cycle long event report that the relay recorded at the time of trip. The voltage and current waveform data is shown in Figure 5.26. Is there a fault on the system? If yes, which type?

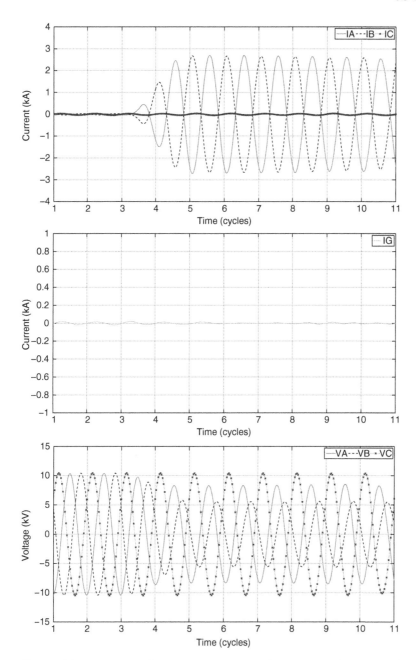

Figure 5.26 Exercise 5.3: Current and voltage waveforms recorded by the relay in the substation at the time of trip.

Solution:

Yes, there is most definitely a line-to-line fault between phases A and B on the distribution feeder. The fault does not involve the ground. In other words, this is not an AB-G fault as the ground current, I_G, is zero. Plus, in a line-to-line fault, the two faulted phases are equal in magnitude and 180 degrees apart as is the case in this event.

■ **Exercise 5.4**

In the previous example, the voltage and current phasors during the fault (cycle 7.5) are given below.

$$I^f_{S_{abc}} [A] \qquad V^f_{S_{abc}} [kV]$$

$$\begin{bmatrix} 1896.30\angle - 117.17° \\ 1876.30\angle 63.35° \\ 24.70\angle 14.04° \end{bmatrix} \quad \begin{bmatrix} 5.9139\angle - 114.13° \\ 3.9371\angle 161.40° \\ 7.3782\angle 35.21° \end{bmatrix}.$$

Voltage and current phasors at the first cycle (prefault) are given below.

$$I^p_{S_{abc}} [A] \qquad V^p_{S_{abc}} [kV]$$

$$\begin{bmatrix} 22.1\angle - 84.81° \\ 17.8\angle 147.65° \\ 26.4\angle 15.35° \end{bmatrix} \quad \begin{bmatrix} 7.3458\angle - 85.79° \\ 7.3618\angle 154.76° \\ 7.3999\angle 34.60° \end{bmatrix}.$$

The 2.65-mile-long distribution feeder has a positive-sequence impedance of $z_1 = 0.7020 + j0.7436$ Ω/mile. Given the above data, determine the fault location using the following methods.

(a) Loop reactance method
(b) Simple reactance method
(c) Takagi method
(d) Modified Takagi method
(e) Girgis et al. method
(f) Santoso et al. method

Solution:

First, we need to calculate the sequence current phasors. Looking at the prefault voltage data at cycle 1, A-phase peaks first in the negative y-axis followed by B-phase and then C-phase. This indicates that the utility has ABC phase rotation. For an AB fault, using the healthy phase (C phase) as the reference, we can calculate the sequence components as:

$$\begin{bmatrix} I^f_{S_0} \\ I^f_{S_1} \\ I^f_{S_2} \end{bmatrix} = \frac{1}{3} \begin{bmatrix} 1 & 1 & 1 \\ 1 & a & a^2 \\ 1 & a^2 & a \end{bmatrix} \times \begin{bmatrix} I^f_{S_c} \\ I^f_{S_a} \\ I^f_{S_b} \end{bmatrix}$$

$$= \frac{1}{3} \begin{bmatrix} 1 & 1 & 1 \\ 1 & a & a^2 \\ 1 & a^2 & a \end{bmatrix} \times \begin{bmatrix} 24.70\angle14.04° \\ 1896.30\angle-117.17° \\ 1876.30\angle63.35° \end{bmatrix}$$

$$= \begin{bmatrix} 1.30\angle-97.10° \\ 1098.20\angle-26.46° \\ 1080.00\angle152.63° \end{bmatrix} \quad [\text{A}].$$

The sequence voltage phasors can be calculated as:

$$\begin{bmatrix} V_{S_0}^f \\ V_{S_1}^f \\ V_{S_2}^f \end{bmatrix} = \frac{1}{3} \begin{bmatrix} 1 & 1 & 1 \\ 1 & a & a^2 \\ 1 & a^2 & a \end{bmatrix} \times \begin{bmatrix} V_{S_c}^f \\ V_{S_a}^f \\ V_{S_b}^f \end{bmatrix}$$

$$= \frac{1}{3} \begin{bmatrix} 1 & 1 & 1 \\ 1 & a & a^2 \\ 1 & a^2 & a \end{bmatrix} \times \begin{bmatrix} 7.3782\angle35.21° \\ 5.9139\angle-114.13° \\ 3.9371\angle161.40° \end{bmatrix}$$

$$= \begin{bmatrix} 0.06\angle137.07° \\ 5.54\angle26.66° \\ 2.06\angle57.21° \end{bmatrix} \quad [\text{kV}].$$

We will also need to calculate voltage V_S and current I_S as given below:

$$V_S = V_{S_a}^f - V_{S_b}^f$$
$$= 5.9139\angle-114.13° - 3.9371\angle161.40°$$
$$= 6.78\angle-78.83° \quad [\text{kV}],$$
$$I_S = I_{S_a}^f - I_{S_b}^f$$
$$= 1896.30\angle-117.17° - 1876.30\angle63.35°$$
$$= 3772.50\angle-116.91° \quad [\text{A}].$$

Now we are ready to solve for the distance to fault.

(a) The loop reactance method uses (5.15) to calculate fault location estimate:

$$m = \frac{\text{imag}\left(\dfrac{V_{S_1}^f - V_{S_2}^f}{I_{S_1}^f}\right)}{\text{imag}(2z_1)}$$

$$= \frac{\text{imag}\left(\dfrac{(5.54\angle26.66° - 2.06\angle57.21°) \times 1000}{1098.20\angle-26.46°}\right)}{(2 \times 0.7436)}$$

$$= 1.46 \quad [\text{mi}].$$

The estimated fault location is 1.46 miles. Line patrol found the fault at 1.42 miles from the substation. A tree limb was the root cause of the fault. After trimming the tree limbs, power was restored after 68 minutes. The estimate from the loop reactance method was very close to the actual location of the fault. Recall that the loop reactance method does not consider load. Because load current was extremely small in this event, around 20 A, the assumption made by the loop reactance method was met, which contributed to the excellent accuracy from this method.

(b) The simple reactance method uses (5.24) to calculate fault location. The calculation steps are shown below:

$$m = \frac{\text{imag}\left(\dfrac{V_S}{I_S}\right)}{\text{imag}(z_1)}$$

$$= \frac{\text{imag}\left(\dfrac{6.78\angle - 78.83° \times 1000}{3772.50\angle - 116.91°}\right)}{0.7436}$$

$$= 1.49 \quad [\text{mi}].$$

(c) The Takagi method accounts for system load by subtracting out the prefault current as shown below:

$$\Delta I_S = \left(I^f_{S_a} - I^p_{S_a}\right) - \left(I^f_{S_b} - I^p_{S_b}\right)$$

$$= \left(1896.30\angle - 117.17° - 22.1\angle - 84.81°\right)$$

$$- \left(1876.30\angle 63.35° - 17.8\angle 147.65°\right)$$

$$= 3752.30\angle - 117.36° \quad [\text{A}].$$

Next, we use (5.25) to calculate fault location as

$$m = \frac{\text{imag}\left(V_S \times \Delta I^*_S\right)}{\text{imag}\left(z_1 \times I_S \times \Delta I^*_S\right)}$$

$$= \frac{\text{imag}\left(6.78\angle - 78.83° \times 1000 \times \left(3752.3\angle - 117.36°\right)^*\right)}{\text{imag}\left((0.702 + j0.7436) \times 3772.5\angle - 116.91° \times \left(3752.3\angle - 117.36°\right)^*\right)}$$

$$= 1.49 \quad [\text{mi}].$$

(d) The modified Takagi method uses I_{seq} to account for load. For an AB fault, it can be calculated as:

$$I_{seq} = jI^f_{S_2}$$

$$= 1080.00\angle - 117.37° \quad [\text{A}].$$

Next, the distance to fault can be calculated as:

$$m = \frac{\text{imag}\left(V_S \times I^*_{seq}\right)}{\text{imag}\left(z_1 \times I_S \times I^*_{seq}\right)}$$

$$= \frac{\text{imag}\left(6.78\angle - 78.83° \times 1000 \times \left(1080.00\angle - 117.37°\right)^*\right)}{\text{imag}\left((0.702 + j0.7436) \times 3772.5\angle - 116.91° \times \left(1080.00\angle - 117.37°\right)^*\right)}$$

$$= 1.49 \quad [\text{mi}].$$

(e) To use the Girgis et al. method, let us first calculate the variables below.

$$R_{app} = \text{real}\left(\frac{V_S}{I_S}\right)$$

$$= \text{real}\left(\frac{6.78\angle - 78.83° \times 1000}{3772.50\angle - 116.91°}\right)$$

$$= 1.4147 \quad \Omega,$$

$$X_{app} = \text{imag}\left(\frac{V_S}{I_S}\right)$$

$$= \text{imag}\left(\frac{6.78\angle - 78.83° \times 1000}{3772.50\angle - 116.91°}\right)$$

$$= 1.1085 \quad \Omega,$$

$$I_d = \text{real}\left(I_{comp}\right) = \text{real}\left(\Delta I_S\right) = -1724.48 \quad [A],$$
$$I_q = \text{imag}\left(I_{comp}\right) = \text{imag}\left(\Delta I_S\right) = -3332.55 \quad [A],$$
$$I_{s1} = \text{real}\left(I_S\right) = \text{real}\left(3772.50\angle - 116.91°\right) = -1707.40 \quad [A],$$
$$I_{s2} = \text{imag}\left(I_S\right) = \text{imag}\left(3772.50\angle - 116.91°\right) = -3364.01 \quad [A],$$

$$L = \frac{I_d I_{s1} + I_q I_{s2}}{I_{s1}^2 + I_{s2}^2}$$

$$= \frac{1724.48 \times 1707.40 + 3332.55 \times 3364.01}{1707.40^2 + 3364.01^2} = 0.9946,$$

$$M = \frac{I_q I_{s1} - I_d I_{s2}}{I_{s1}^2 + I_{s2}^2}$$

$$= \frac{3332.55 \times 1707.40 - 1724.48 \times 3364.01}{1707.40^2 + 3364.01^2} = -0.0078.$$

The distance to fault can then be calculated as:

$$m = \frac{X_{app}L - R_{app}M}{x_1 L - r_1 M}$$

$$= \frac{1.1085 \times 0.9946 - 1.4147 \times (-0.0078)}{0.7436 \times 0.9946 - 0.702 \times (-0.0078)}$$

$$= 1.49 \quad [\text{mi}].$$

The fault resistance is estimated to be 0.69 Ω as shown below:

$$R_f = \frac{R_{app} - mr_1}{L}$$

$$= \frac{1.4147 - 1.49 \times 0.702}{0.9946}$$

$$= 0.37 \quad \Omega.$$

(f) When using the Santoso et al. method, the first step is to calculate variable p_1 using (5.37) as:

$$p_1 = \frac{I_{comp}}{I_S} = \frac{\Delta I_S}{I_S}$$

$$= \frac{3752.30\angle - 117.36°}{3772.50\angle - 116.91°}$$

$$= 0.9946 - j0.0078,$$

$$\therefore u_1 = \text{real}\,(p_1) = 0.9946$$

$$\text{and} \quad v_1 = \text{imag}\,(p_1) = -0.0078.$$

The distance to fault can then be solved using (5.42) as

$$|m| = \frac{u_1 X_{app} - v_1 R_{app}}{|z_1||p_1|}$$

$$= \frac{0.9946 \times 1.1085 + 0.0078 \times 1.4147}{1.0226 \times 0.9946}$$

$$= 1.09 \quad [\text{mi}].$$

Remember that the Santoso et al. method provides the lower bound of the distance estimate. Line patrol should start their search from 1.09 miles and move further down the feeder till they find the fault. All the other methods performed exceptionally well and were extremely close to the actual fault location of 1.42 miles. The Girgis method estimated the fault to have a resistance of 0.37 Ω.

■ Exercise 5.5

A relay protecting a distribution feeder that has a nominal line-to-ground voltage of 4.8 kV issued a trip and opened up the circuit breaker at the substation. This particular feeder serves some major customers including a national bank and a racquet ball club. As a result, it was imperative that the utility find the fault and restore power as quickly as possible. The relay that had issued a trip had also recorded a 15-cycle event report shown in Figure 5.27. You can see that the distribution feeder had experienced a three-phase fault. Currents on all three phases were extremely high and line-to-ground voltages on all three phases had sagged. When the circuit breaker opened up and cleared the fault, the three-phase voltages returned back to nominal. This event report did not contain prefault information.

A fast Fourier transform filter was used to extract the fault voltage, and current phasors at the second cycle and are given below:

$$
I^f_{S_{abc}}\ [\text{A}] \qquad\qquad V^f_{S_{abc}}\ [\text{kV}]
$$

$$
\begin{bmatrix} 2331.30\angle 170.99° \\ 2386.20\angle 42.78° \\ 2061.40\angle -74.45° \end{bmatrix}
\begin{bmatrix} 3.08\angle -146.52° \\ 2.97\angle 96.04° \\ 3.13\angle -23.38° \end{bmatrix}.
$$

The distribution feeder is 1.57 miles long. The positive-sequence line impedance setting in the relay is $z_1 = 0.3084 + j0.6054\ \Omega/\text{mile}$. Given the above data, determine the fault location using the following methods.

(a) Loop reactance method
(b) Simple reactance method.

Solution:

(a) The loop reactance method uses (5.22) to calculate the distance to a three-phase fault. Any one of the three phases can be used for fault location calculations. In this example, we use the A-phase.

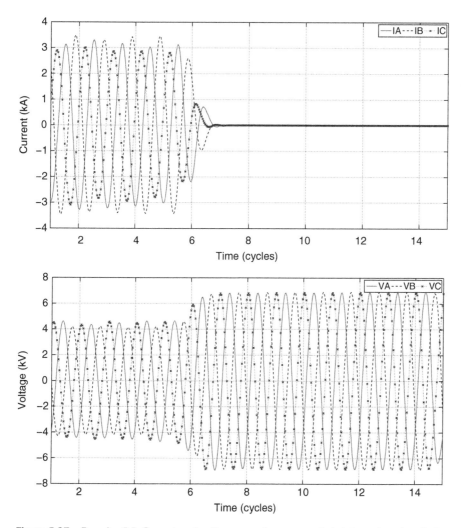

Figure 5.27 Exercise 5.5: Current and voltage waveforms recorded at the substation during a three-phase fault.

$$
\begin{aligned}
m &= \frac{\text{imag}\left(\dfrac{V_{phase}}{I_{phase}}\right)}{\text{imag}\left(z_1\right)} \\[2ex]
&= \frac{\text{imag}\left(\dfrac{V_{S_a}^f}{I_{S_a}^f}\right)}{\text{imag}\left(z_1\right)} \\[2ex]
&= \frac{\text{imag}\left(\dfrac{3.08\angle -146.52° \times 1000}{2331.30\angle 170.99°}\right)}{0.6054} \\[2ex]
&= 1.47 \quad [\text{mi}].
\end{aligned}
$$

(b) To implement the simple reactance method, we need to first calculate V_S and I_S. For a three-phase fault, they can be calculated as shown below per Table 5.1.

$$V_S = V_{S_a}^f - V_{S_b}^f$$
$$= 3.08\angle - 146.52° - 2.97\angle 96.04°$$
$$= 5.17\angle - 115.88° \quad [\text{kV}],$$
$$I_S = I_{S_a}^f - I_{S_b}^f$$
$$= 2331.30\angle 170.99° - 2386.20\angle 42.78°$$
$$= 4244.00\angle - 162.79° \quad [\text{A}].$$

Next we can calculate the distance to fault using (5.24):

$$m = \frac{\text{imag}\left(\dfrac{V_S}{I_S}\right)}{\text{imag}(z_1)}$$

$$= \frac{\text{imag}\left(\dfrac{5.17\angle - 115.88° \times 1000}{4244.00\angle - 162.79°}\right)}{0.6054}$$

$$= 1.47 \quad [\text{mi}].$$

The utility found the fault at 1.51 miles from the substation. Strong winds had caused a tree branch to fall on the line and create a fault. After trimming the tree and removing some of the tree limbs, power was restored back to the customers after 29 minutes. You can see that both the loop reactance and simple reactance methods performed well and were close to the actual fault location. Note that we could not apply the Takagi, Girgis et al., Santoso et al., and Novosel methods because the prefault voltage and current information were not available. The modified Takagi method does not apply to three-phase faults.

■ Exercise 5.6

One fine Sunday evening, a customer called in to report a momentary blink (notice how faults have a tendency to occur at the most inopportune times!). Weather was fair at the time of the blink. A lineman was dispatched to investigate the root cause of the blink. The utility would like to help the lineman by telling him where to go and search for the fault so that he can find the fault faster and get back to enjoying whatever is left of his Sunday. They were able to communicate with a relay at the substation that had recorded a 16-cycle long event report during the fault. This is shown in Figure 5.28. By looking at the current and voltage waveforms, you can tell that the distribution feeder had indeed experienced a fault. The fault started off as a C-G fault and then evolved into a CA-G fault. The presence of ground current plus the offset between the two faulted phases (they are not 180 degrees out of phase with each other) confirms that this is a CA-G fault and not a CA fault.

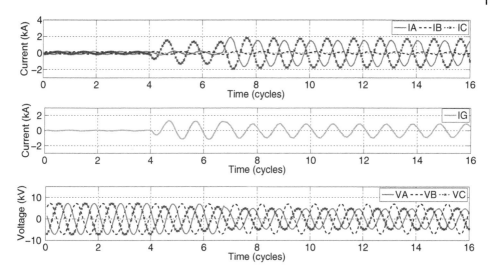

Figure 5.28 Exercise 5.6: Current and voltage waveforms recorded at the substation at the time of the momentary blink. A C-G fault evolved into a CA-G fault. The two faulted phases are not exactly 180 degrees apart. Plus the presence of ground current confirms that the fault was indeed a CA-G fault and not a CA fault.

A fast Fourier transform filter was applied to cycle 12.5 to extract the fault voltage and current phasors. They are given below:

$$
\begin{matrix}
I^f_{S_{abc}} \text{ [A]} & V^f_{S_{abc}} \text{ [kV]} \\
\begin{bmatrix} 1588.90\angle - 156.23° \\ 166.90\angle 133.68° \\ 1768.70\angle - 10.68° \end{bmatrix} & \begin{bmatrix} 4.80\angle - 74.85° \\ 7.05\angle 162.72° \\ 4.95\angle 27.44° \end{bmatrix}.
\end{matrix}
$$

Prefault voltage and current phasors, calculated at cycle 0.5 cycle, are given below:

$$
\begin{matrix}
I^p_{S_{abc}} \text{ [A]} & V^p_{S_{abc}} \text{ [kV]} \\
\begin{bmatrix} 181.50\angle - 96.12° \\ 143.50\angle 147.25° \\ 136.50\angle 27.84° \end{bmatrix} & \begin{bmatrix} 7.3122\angle - 79.39° \\ 7.3206\angle 161.32° \\ 7.3771\angle 40.87° \end{bmatrix}.
\end{matrix}
$$

The distribution feeder has a positive-sequence line impedance of $z_1 = 0.2780 + j0.6584$ Ω/mile. Determine fault location using the following methods:

(a) Loop reactance method
(b) Simple reactance method
(c) Takagi method
(d) Modified Takagi method
(e) Girgis et al. method
(f) Santoso et al. method

Solution:

Before diving into fault location calculations, we need to calculate the sequence current phasors. From the prefault data, it is evident that the utility has ABC phase rotation. Since this is a CA-G fault, we use the healthy phase (B phase) as the reference when calculating the sequence components as shown below:

$$
\begin{bmatrix} I^f_{S_0} \\ I^f_{S_1} \\ I^f_{S_2} \end{bmatrix} = \frac{1}{3} \begin{bmatrix} 1 & 1 & 1 \\ 1 & a & a^2 \\ 1 & a^2 & a \end{bmatrix} \times \begin{bmatrix} I^f_{S_b} \\ I^f_{S_c} \\ I^f_{S_a} \end{bmatrix}
$$

$$
= \frac{1}{3} \begin{bmatrix} 1 & 1 & 1 \\ 1 & a & a^2 \\ 1 & a^2 & a \end{bmatrix} \times \begin{bmatrix} 166.90\angle 133.68° \\ 1768.70\angle -10.68° \\ 1588.90\angle -156.23° \end{bmatrix}
$$

$$
= \begin{bmatrix} 288.00\angle -78.75° \\ 1136.80\angle 98.90° \\ 719.90\angle -89.65° \end{bmatrix} \quad [A].
$$

The sequence voltage phasors can be calculated as:

$$
\begin{bmatrix} V^f_{S_0} \\ V^f_{S_1} \\ V^f_{S_2} \end{bmatrix} = \frac{1}{3} \begin{bmatrix} 1 & 1 & 1 \\ 1 & a & a^2 \\ 1 & a^2 & a \end{bmatrix} \times \begin{bmatrix} V^f_{S_b} \\ V^f_{S_c} \\ V^f_{S_a} \end{bmatrix}
$$

$$
= \frac{1}{3} \begin{bmatrix} 1 & 1 & 1 \\ 1 & a & a^2 \\ 1 & a^2 & a \end{bmatrix} \times \begin{bmatrix} 7.05\angle 162.72° \\ 4.95\angle 27.44° \\ 4.80\angle -74.85° \end{bmatrix}
$$

$$
= \begin{bmatrix} 0.37\angle -166.81° \\ 5.55\angle 158.93° \\ 1.20\angle 171.28° \end{bmatrix} \quad [kV].
$$

We will also need to calculate voltage V_S and current I_S as given below.

$$
V_S = V^f_{S_c} - V^f_{S_a}
$$
$$
= 4.95\angle 27.44° - 4.80\angle -74.85°
$$
$$
= 7.59\angle 65.58° \quad [kV],
$$
$$
I_S = I^f_{S_c} - I^f_{S_a}
$$
$$
= 1768.70\angle -10.68° - 1588.90\angle -156.23°
$$
$$
= 3207.45\angle 5.59° \quad [A].
$$

Now we are ready to solve for the distance to fault.

(a) The loop reactance method uses (5.19) to calculate the distance to a double line-to-ground fault. Calculations are shown below.

$$m = \frac{\text{imag}\left(\dfrac{V_{S_1}^f - V_{S_2}^f}{I_{S_1}^f - I_{S_2}^f}\right)}{\text{imag}(z_1)}$$

$$= \frac{\text{imag}\left(\dfrac{(5.55\angle158.93° - 1.20\angle171.28°) \times 1000}{1136.80\angle98.90° - 719.90\angle -89.65°}\right)}{0.6584}$$

$$= 3.11 \quad [\text{mi}].$$

(b) Distance to fault using the simple reactance method can be calculated using (5.24):

$$m = \frac{\text{imag}\left(\dfrac{V_S}{I_S}\right)}{\text{imag}(z_1)}$$

$$= \frac{\text{imag}\left(\dfrac{7.59\angle65.58° \times 1000}{3207.45\angle5.59°}\right)}{0.6584}$$

$$= 3.11 \quad [\text{mi}].$$

(c) The Takagi method accounts for system load by subtracting out the prefault current as shown below.

$$\Delta I_S = \left(I_{S_c}^f - I_{S_c}^p\right) - \left(I_{S_a}^f - I_{S_a}^p\right)$$

$$= (1768.70\angle -10.68° - 136.50\angle27.84°)$$

$$\quad - (1588.90\angle -156.23° - 181.50\angle -96.12°)$$

$$= 3052.90\angle1.28° \quad [\text{A}].$$

Next, we use (5.25) to calculate fault location as

$$m = \frac{\text{imag}\left(V_S \times \Delta I_S^*\right)}{\text{imag}\left(z_1 \times I_S \times \Delta I_S^*\right)}$$

$$= \frac{\text{imag}\left(7.59\angle65.58° \times 1000 \times \left(3052.9\angle1.28°\right)^*\right)}{\text{imag}\left((0.2780 + j0.6584) \times 3207.45\angle5.59° \times \left(3052.9\angle1.28°\right)^*\right)}$$

$$= 3.15 \quad [\text{mi}].$$

(d) The modified Takagi method uses I_{seq} to account for system load. It can be calculated as:

$$I_{seq} = jI_{S_2}^f$$

$$= 719.90\angle0.35° \quad [\text{A}].$$

Next, the distance to fault can be calculated as:

$$m = \frac{\text{imag}\left(V_S \times I_{seq}^*\right)}{\text{imag}\left(z_1 \times I_S \times I_{seq}^*\right)}$$

$$= \frac{\text{imag}\left(7.59\angle 65.58° \times 1000 \times \left(719.9\angle 0.35°\right)^{*}\right)}{\text{imag}\left(\left(0.2780 + j0.6584\right) \times 3207.45\angle 5.59° \times \left(719.9\angle 0.35°\right)^{*}\right)}$$

$$= 3.16 \quad [\text{miles}].$$

(e) To use the Girgis et al. method, let us first calculate the variables below.

$$R_{app} = \text{real}\left(\frac{V_S}{I_S}\right)$$

$$= \text{real}\left(\frac{7.59\angle 65.58° \times 1000}{3207.45\angle 5.59°}\right)$$

$$= 1.1835 \quad \Omega,$$

$$X_{app} = \text{imag}\left(\frac{V_S}{I_S}\right)$$

$$= \text{imag}\left(\frac{7.59\angle 65.58° \times 1000}{3207.45\angle 5.59°}\right)$$

$$= 2.0491 \quad \Omega,$$

$$I_d = \text{real}\left(I_{comp}\right) = \text{real}\left(\Delta I_S\right) = 3052.14 \quad [\text{A}],$$
$$I_q = \text{imag}\left(I_{comp}\right) = \text{imag}\left(\Delta I_S\right) = 68.20 \quad [\text{A}],$$
$$I_{s1} = \text{real}\left(I_S\right) = \text{real}\left(3207.45\angle 5.59°\right) = 3192.20 \quad [\text{A}],$$
$$I_{s2} = \text{imag}\left(I_S\right) = \text{imag}\left(3207.45\angle 5.59°\right) = 312.44 \quad [\text{A}],$$

$$L = \frac{I_d I_{s1} + I_q I_{s2}}{I_{s1}^2 + I_{s2}^2}$$

$$= \frac{3052.14 \times 3192.20 + 68.20 \times 312.44}{3192.20^2 + 312.44^2} = 0.9491,$$

$$M = \frac{I_q I_{s1} - I_d I_{s2}}{I_{s1}^2 + I_{s2}^2}$$

$$= \frac{68.20 \times 3192.20 - 3052.14 \times 312.44}{3192.20^2 + 312.44^2} = -0.0715.$$

The distance to fault can then be calculated as:

$$m = \frac{X_{app}L - R_{app}M}{x_1 L - r_1 M}$$

$$= \frac{2.0491 \times 0.9491 - 1.1835 \times (-0.0715)}{0.6584 \times 0.9491 - 0.2780 \times (-0.0715)}$$

$$= 3.15 \quad [\text{miles}].$$

The fault resistance is estimated to be 0.33 Ω as shown below.

$$R_f = \frac{R_{app} - mr_1}{L}$$

$$= \frac{1.1835 - 3.15 \times 0.2780}{0.9491}$$

$$= 0.33 \quad \Omega.$$

(f) The Santoso et al. method calculates the lower bound of the distance estimate. When using this method, the first step is to calculate variable p_1 using (5.37) as:

$$p_1 = \frac{I_{comp}}{I_S} = \frac{\Delta I_S}{I_S}$$

$$= \frac{3052.90 \angle 1.28°}{3207.45 \angle 5.59°}$$

$$= 0.9491 - j0.0715,$$

$$\therefore u_1 = \text{real}\,(p_1) = 0.9491$$

$$\text{and} \quad v_1 = \text{imag}\,(p_1) = -0.0715.$$

The distance to fault can then be solved using (5.42) as

$$|m| = \frac{u_1 X_{app} - v_1 R_{app}}{|z_1||p_1|}$$

$$= \frac{0.9491 \times 2.0491 + 0.0715 \times 1.1835}{0.7147 \times 0.9518}$$

$$= 2.99 \quad [\text{miles}].$$

Combining results from all the fault location algorithms, the fault is expected to be between 2.99 and 3.16 miles. It is also estimated that the fault has a resistance of 0.33 Ω. The lineman found two barrels on two cutouts on a capacitor bank to be completely burned at a distance of 3.10 miles from the substation. One of the barrels was completely burned into two with a portion of the barrel on the ground. The reason that caused the cutouts to flash was not obvious. The lineman returned back to the operations center and put in a request to check the capacitor bank. The event did not result in an outage. Customers saw a voltage sag and a blink.

5.4 Summary

In this chapter, we present the strengths, weaknesses, and data requirements of several impedance-based fault location algorithms that can be used to locate faults on overhead distribution feeders. We discuss several challenges associated with locating faults on distribution feeders and present solutions to overcome some of those challenges. We then work through several exercises to complement the theory presented in this chapter.

6

Distribution Fault Location With Current Only

Impedance-based fault location algorithms require both voltage and current phasors to compute the distance to a fault. Unfortunately, this requirement cannot always be met in distribution feeders because they are protected by overcurrent relays or recloser controls that only use current measurements to detect and isolate faults. Voltage measurements are thus optional in such relays and not always available for fault location. Voltage measurements can also be missing when a fuse protecting the potential transformer blows and results in a loss of potential [105]. In such scenarios, existing impedance-based algorithms cannot be used for fault location. This chapter develops fault location algorithms that use current data as the only input for calculating fault location. These current-only algorithms complement existing fault location algorithms and allow system operators to perform fault location even in the absence of voltage data. The algorithms are developed in two parts: fault location using current phasors (both magnitude and phase angle available) and fault location using current magnitude (no phase angle available). The algorithms use the source impedance data and Kirchhoff's circuit laws to estimate the missing fault voltage at the relay location [106]. Once the fault voltage is available, impedance-based fault location principles are applied from the relay location to estimate the distance to fault. Another approach searches for the fault location by matching the short-circuit current in the system short-circuit model with the measured fault current [107, 108]. The methods listed in the chapter are non-iterative and straightforward to implement.

6.1 Current Phasors Only Method

A radial distribution feeder l miles long is shown in Figure 6.1. The feeder is fed by an interconnected transmission system through a substation transformer. The feeder is homogeneous and has a positive- and zero-sequence line impedance of z_1 and z_0 ohms per mile, respectively. All loads served by the distribution feeder are lumped at the end of the feeder and represented by an impedance of Z_{load} ohms. When a single line-to-ground fault occurs at m miles from node A with a fault resistance of R_f ohms, a digital relay at node A records the current waveform before and during the fault. From the current waveform, it is possible to extract the prefault current phasors, $I_{A_{abc}}$, and fault current phasors, $I^f_{A_{abc}}$, in units of amperes. Phase voltages at node A during the fault, $V^f_{A_{abc}}$, are not available. Note that

Fault Location on Transmission and Distribution Lines: Principles and Applications, First Edition.
Swagata Das, Surya Santoso, and Sundaravaradan N. Ananthan.
© 2022 John Wiley & Sons Ltd. Published 2022 by John Wiley & Sons Ltd.
Companion website: www.wiley.com/go/das/faultlocation

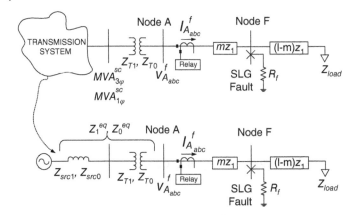

Figure 6.1 Transmission network upstream from the distribution transformer represented by an equivalent Thevenin circuit.

the superscript f indicates that the corresponding measurement has been obtained during a fault. The method outlined below shows how to calculate fault location with only the current phasors.

Step 1: Estimate Source Impedance Behind Relay

For the purpose of calculating fault location, it is necessary to represent the upstream transmission circuit by a Thevenin circuit, which consists of an ideal voltage source in series with an equivalent impedance. This impedance, often referred to as the system short-circuit impedance, can be obtained from the three-phase ($MVA_{3\varphi}^{sc}$) and single-phase ($MVA_{1\varphi}^{sc}$) short-circuit capacities on the high side of the distribution transformer [89]. These short-circuit capacities are usually obtained from the short-circuit model and are given as phasors. However, if no phase angle information is provided, assume that the equivalent impedance is a pure reactance. The positive-sequence short-circuit impedance can be obtained as:

$$Z_{src1} = \frac{kV_{LL}^2}{MVA_{3\phi}^{sc*}} \quad [\Omega], \tag{6.1}$$

where kV_{LL} is the rated line-to-line voltage in kilovolts. The zero-sequence short-circuit impedance can be obtained as

$$Z_{src0} = \frac{3 \times kV_{LL}^2}{MVA_{1\phi}^{sc*}} - 2Z_{src1} \quad [\Omega]. \tag{6.2}$$

Once the short-circuit impedances are known, the equivalent positive- and zero-sequence source impedances, Z_1^{eq} and Z_0^{eq}, behind the relay can be calculated as

$$Z_1^{eq} = Z_{scr1} + Z_{T1} \quad [\Omega],$$
$$Z_0^{eq} = Z_{scr0} + Z_{T0} \quad [\Omega] \quad \text{(Depends of the transformer construction)}, \tag{6.3}$$

where Z_{T1} and Z_{T0} are the positive- and zero-sequence impedances of the distribution transformer. Z_1^{eq} and Z_0^{eq} will be used in the next step to estimate fault voltage at node A. It is important to keep in mind that the short-circuit impedance of the transmission network may not always remain constant as generators may come online and go offline at different times of the day. Fortunately, the impedance of the distribution transformer is constant and much greater than the short-circuit impedance of the transmission network. Therefore, for all practical purposes, the source impedance behind the relay can be assumed to be constant. In fact, if the short-circuit capacities are not available, set the source impedance behind the relay equal to the distribution transformer impedance. Also, when calculating Z_0^{eq}, pay attention to the transformer connection. If the transformer connection is delta-wye, Z_0^{eq} will equal Z_{T0}.

Step 2: Estimate Sequence Components of the Missing Fault Voltage

The next task is to estimate the sequence components of the fault voltage phasor at node A. In the previous chapter, we assumed that this voltage was measured by a potential transformer and was available for calculating fault location. In this section, we attempt to estimate the sequence components of this missing voltage phasor.

Figure 6.2 shows the connection of the sequence networks during a single line-to-ground fault. Notations in the figure are defined as follows: $V_{A_1}^f$, $V_{A_2}^f$, and $V_{A_0}^f$ are the positive-,

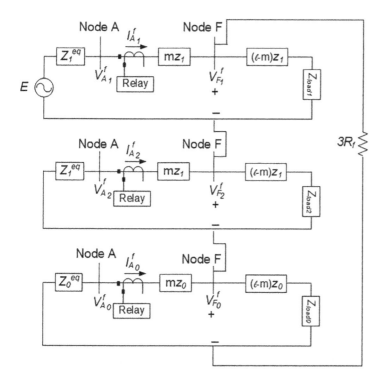

Figure 6.2 Interconnection of sequence networks during a single line-to-ground fault on a distribution feeder.

negative-, and zero-sequence voltage phasors at node A during the fault in volts, respectively, $I^f_{A_1}$, $I^f_{A_2}$, and $I^f_{A_0}$ are the positive-, negative-, and zero-sequence current phasors measured by the relay during the fault in amperes, respectively, Z_{load1}, Z_{load2}, and Z_{load0} are the positive-, negative-, and zero-sequence impedances of the load in ohms, respectively, $V^f_{F_1}$, $V^f_{F_2}$ and $V^f_{F_0}$ are the positive-, negative-, and zero-sequence voltage phasors at the fault point (node F) in volts, respectively, and I_F is the current phasor at the fault point (node F) in amperes. Our goal is to estimate $V^f_{A_1}$, $V^f_{A_2}$, and $V^f_{A_0}$. Voltage phasor $V^f_{A_2}$ can be estimated using Kirchhoff's circuit laws as

$$V^f_{A_2} = -\left(Z^{eq}_1 \times I^f_{A_2} \right) \quad [\text{V}]. \tag{6.4}$$

In a similar manner, $V^f_{A_0}$ can be estimated as

$$V^f_{A_0} = -\left(Z^{eq}_0 \times I^f_{A_0} \right) \quad [\text{V}]. \tag{6.5}$$

The calculation of $V^f_{A_1}$, on the other hand, is complicated by the presence of the internal generator voltage source, E. To develop a workaround, we use the superposition principle to decompose the network during fault into a prefault and pure fault network, which is illustrated in Figure 6.3. The concept of a pure fault network has been described in Section 3.1.2. Notations in the figure are defined as follows: V_{A_1} is the positive-sequence voltage phasor at node A before fault in volts, I_{A_1} is the positive-sequence current phasor measured by the relay before fault in amperes, V_{F_1} is the positive-sequence voltage phasor at node F before fault in volts, $\Delta V^f_{A_1}$, $\Delta V^f_{A_2}$, and $\Delta V^f_{A_0}$ are the positive-, negative-, and zero-sequence pure fault voltage phasors at node A in volts, respectively, $\Delta V^f_{F_1}$, $\Delta V^f_{F_2}$, and $\Delta V^f_{F_0}$ are the positive-, negative-, and zero-sequence pure fault voltage phasors at node F in volts, respectively, and $\Delta I^f_{A_1}$, $\Delta I^f_{A_2}$, and $\Delta I^f_{A_0}$ are the positive-, negative-, and zero-sequence pure fault current phasors measured by the relay in amperes, respectively. Fault voltage phasor at node A, $V^f_{A_1}$, is the summation of the prefault voltage phasor and the voltage that exists only during the fault:

$$V^f_{A_1} = V_{A_1} + \Delta V^f_{A_1} \quad [\text{V}] \tag{6.6}$$

Now, the magnitude of V_{A_1} is close to 1 per unit in any practical power system. The phase angle of V_{A_1} ($\theta_{v_{a1}}$) can be obtained from the power factor (pf) as

$$\theta_{v_{a1}} = \cos^{-1}(\text{pf}) + \theta_{i_{a1}} \quad [\text{degrees}], \tag{6.7}$$

where $\theta_{i_{a1}}$ is the phase angle of the prefault current, I_{A_1}. Note that the power factor can be determined by running a load flow analysis on the circuit model of the distribution feeder. The pure fault voltage phasor can be estimated as

$$\Delta V^f_{A_1} = -\left(Z^{eq}_1 \times \Delta I^f_{A_1} \right) = -\left[Z^{eq}_1 \times \left(I^f_{A_1} - I_{A_1} \right) \right] \quad [\text{V}]. \tag{6.8}$$

Step 3: Transform Fault Voltage from Sequence to Phasor Domain

Because impedance-based fault location algorithms use phase voltages and not sequence components to calculate fault location, this step transforms the estimated fault voltages at

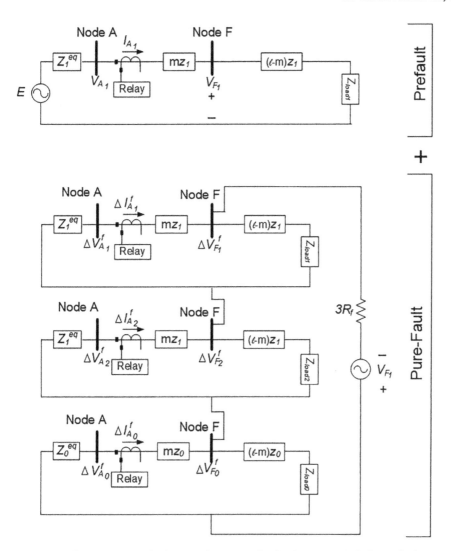

Figure 6.3 Superposition principle applied to the distribution network during a single line-to-ground fault.

the relay location from the sequence to the phasor domain. An example transformation for an A-G fault and ABC system phase rotation is shown below.

$$
\begin{bmatrix} V^f_{A_a} \\ V^f_{A_b} \\ V^f_{A_c} \end{bmatrix} = \begin{bmatrix} 1 & 1 & 1 \\ 1 & a^2 & a \\ 1 & a & a^2 \end{bmatrix} \times \begin{bmatrix} V^f_{A_0} \\ V^f_{A_1} \\ V^f_{A_2} \end{bmatrix} \quad [V]; \quad a = 1\angle120°. \tag{6.9}
$$

Step 4: Calculate Fault Location

Once the missing phase fault voltage phasor is available, apply existing impedance-based fault location algorithms to estimate the fault location. As an example, the simple reactance method can be used to estimate the distance to an A-G fault as shown below:

$$m = \frac{\text{imag}\left(\dfrac{V_{A_a}^f}{I_{A_a}^f + kI_{A_0}^f}\right)}{\text{imag}\left(z_1\right)} \quad \text{[miles]}. \tag{6.10}$$

The derivation in this section shows that it is possible to locate single line-to-ground faults with current phasors (voltage phasors not available) in four simple steps. The same approach can also be extended to locate all other fault types. Data required is minimal and consists of source impedance behind the relay, power factor, and current phasors measured by the relay before and during the fault.

6.2 Current Magnitude Only Method

Consider the same distribution feeder shown in Figure 6.1. In the event of a fault, suppose that only the fault current magnitudes in phases A, B, and C, measured by an IED at the substation, are known. The phase angle relationship between the three phase currents is not available. Magnitude of the voltage sag during the fault is also not available. In this section, we explore how to locate three-phase, line-to-line, and single line-to-ground faults with such limited data.

(a) Three-Phase Fault

Assume that a three-phase fault occurs at a distance m from node A. The only data available for fault location are the fault current magnitudes, $|I_{A_a}^f|$, $|I_{A_b}^f|$, and $|I_{A_c}^f|$, measured and recorded by the relay at node A. We start by applying the superposition principle and decomposing the network during fault into prefault and pure fault networks as shown in Figure 6.4. Because this is a three-phase fault, we draw only the positive-sequence network. The missing positive-sequence voltage phasor at node A during the fault, $V_{A_1}^f$, is the summation of the prefault voltage phasor and the pure fault voltage phasor (derived in the earlier section) as shown below:

$$V_{A_1}^f = V_{A_1} + \Delta V_{A_1}^f = V_{A_1} - \left[Z_1^{eq} \times \left(I_{A_1}^f - I_{A_1}\right)\right] \tag{6.11}$$

For a three-phase balanced fault, the positive-sequence fault current phasor at node A, $I_{A_1}^f$, is nothing but one of the phase current phasors, phase A in this derivation. In a similar manner, the positive-sequence voltage phasors at node A before and during the fault, V_{A_1} and $V_{A_1}^f$, are nothing but one of the phase voltage phasors. We choose phase A to be consistent in our derivation. Furthermore, we neglect the prefault current from the fault location calculation. This is a reasonable assumption to make since fault current during a three-phase fault is many times greater than the prefault current. As a result, (6.11) simplifies to:

$$V_{A_a}^f = V_{A_a} - \left(Z_1^{eq} \times I_{A_a}^f\right). \tag{6.12}$$

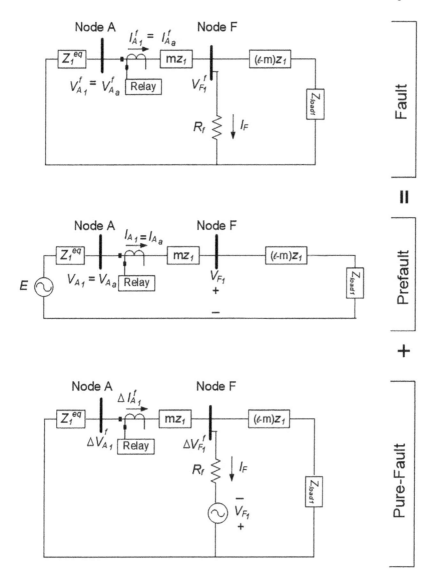

Figure 6.4 Superposition principle applied to the distribution network during a three-phase fault.

Rewriting (6.12) in polar form, we get:

$$|V^f_{A_a}|\angle\theta^f_{v_a} = |V_{A_a}|\angle\theta_{v_a} - |Z^{eq}_1| \times |I^f_{A_a}|\angle\left(\theta_z + \theta^f_{i_a}\right),\qquad(6.13)$$

where $\theta^f_{v_a}$ is the phase angle of the A-phase voltage at node A during fault, θ_{v_a} is the phase angle of the A-phase voltage at node A prior to the fault, θ_z is the phase angle of Z^{eq}_1, and $\theta^f_{i_a}$ is the phase angle of A-phase fault current, $I^f_{A_a}$. Solving for $|V^f_{A_a}|$ from (6.13) appears to be an uphill task as we have one equation but four unknowns: $|V^f_{A_a}|$, $\theta^f_{v_a}$, θ_{v_a}, and $\theta^f_{i_a}$. To reduce the number of unknowns, we use the reverse triangular inequality theorem [109] to calculate $|V_{A_a}^f|$ as:

$$\left|V_{A_a}^f\right| \geq \left||V_{A_a}| - \left(|Z_1^{eq}| \times |I_{A_a}^f|\right)\right| \qquad \because |u - v| \geq ||u| - |v||. \tag{6.14}$$

The equality condition will be valid only when $\theta_{v_a} = \theta_z + \theta_{i_a}^f$. Assuming this condition to hold true, $|V_{A_a}^f|$ can be estimated as

$$\left|V_{A_a}^f\right| = \left||V_{A_a}| - \left(|Z_1^{eq}| \times |I_{A_a}^f|\right)\right|. \tag{6.15}$$

Next, we calculate fault location from the fault network shown in Figure 6.4 by writing the voltage drop equation from node A as:

$$V_{A_a}^f - m \times z_1 \times I_{A_a}^f = I_F \times R_f. \tag{6.16}$$

Assuming a bolted fault, the distance to fault is given by the equation below.

$$m = \frac{V_{A_a}^f}{z_1 \times I_{A_a}^f}. \tag{6.17}$$

Rewriting (6.17) in polar form,

$$m = \frac{|V_{A_a}^f|}{|z_1| \times |I_{A_a}^f|} \angle \left(\theta_{v_a}^f - \theta_{z_{l1}} - \theta_{i_a}^f\right), \tag{6.18}$$

where $\theta_{z_{l1}}$ is the phase angle of the positive-sequence line impedance. Since m is a real number, it can be calculated from only the magnitude terms as

$$m = \frac{|V_{A_a}^f|}{|z_1| \times |I_{A_a}^f|}. \tag{6.19}$$

In the above equation, $|V_{A_a}^f|$ is to be estimated using (6.15), $|z_1|$ is the magnitude of the positive-sequence line impedance in ohms per mile (known), and $|I_{A_a}^f|$ is the magnitude of the fault current recorded by the relay in phase A during a three-phase fault. The equation is simple and straightforward to implement and assumes a bolted fault and a lightly loaded feeder. Another assumption in the calculation of $|V_{A_a}^f|$ is that θ_{v_a} is equal to the summation of θ_z and $\theta_{i_a}^f$. When this assumption is not met, the estimated value of $|V_{A_a}^f|$ will be less than its actual value. As a result, (6.19) will underestimate the distance to the location of the fault. Put another way, the distance estimate using (6.19) can be regarded as a lower bound for the actual distance to the fault. Exercise 6 at the end of the chapter shows that the error due to this angle mismatch is not substantial and that the location estimate computed using this method is close to the actual fault location.

(b) Line-to-Line Fault

Suppose that a line-to-line fault occurs between phase A and phase B of a distribution feeder. A relay at the substation records only the magnitude of currents in the faulted phases.

Voltage in the faulted phases or phase angle relationship between the currents is not available. We begin the task of fault location by reconstructing the fault current phasors from the fault current magnitudes. This step is necessary as fault location will be computed in the sequence domain, and the transformation from the phasor to the sequence domain requires the use of phasors. Assign a phase angle value (50 degrees in this example) to the phase A fault current, $I^f_{A_a}$. For an AB fault, the fault current in phase B is equal but opposite to that in phase A as described in Chapter 2. Finally, neglect the load current in the healthy phase, $I^f_{A_c}$. This is a reasonable assumption since the currents in the faulted phases are many times greater than the load current. The reconstructed fault current phasors are given as follows:

$$I^f_{A_a} = |I^f_{A_a}| \times (\cos\ 50° + j\sin\ 50°) \quad [A],$$
$$I^f_{A_b} = -I^f_{A_a} \quad [A], \tag{6.20}$$
$$I^f_{A_c} = 0 \quad [A].$$

The second step is to estimate the sequence current phasors. Because this is an A-B fault, phase C needs to be used as the reference when calculating sequence components as shown below. System phase rotation is assumed to be ABC.

$$\begin{bmatrix} I^f_{A_0} \\ I^f_{A_1} \\ I^f_{A_2} \end{bmatrix} = \frac{1}{3} \begin{bmatrix} 1 & 1 & 1 \\ 1 & a & a^2 \\ 1 & a^2 & a \end{bmatrix} \times \begin{bmatrix} I^f_{A_c} \\ I^f_{A_a} \\ I^f_{A_b} \end{bmatrix}; \quad a = 1\angle120° \tag{6.21}$$

$$\therefore I^f_{A_0} = 0; \quad I^f_{A_1} = -I^f_{A_2} = \frac{1}{3} I^f_{A_a} (a - a^2) \quad [A].$$

The third step is to estimate the distance to a line-to-line fault. For this purpose, the superposition principle is applied to the distribution network during the line-to-line fault and is shown in Figure 6.5. Notice that the positive- and negative-sequence fault voltage phasors at the fault point are related as shown below:

$$V^f_{F_1} - V^f_{F_2} = I_F \times R_f \tag{6.22}$$

The positive-sequence voltage phasor at the fault point, $V^f_{F_1}$, can be expressed as:

$$\begin{aligned} V^f_{F_1} &= V^f_{A_1} - mz_1 \times I^f_{A_1} \\ &= V_{A_1} + \Delta V^f_{A_1} - mz_1 \times I^f_{A_1} \\ &= V_{A_1} - Z^{eq}_1 \left(I^f_{A_1} - I_{A_1} \right) - mz_1 \times I^f_{A_1}. \end{aligned} \tag{6.23}$$

Neglecting prefault current, the equation above simplifies to:

$$V^f_{F_1} = V_{A_1} - Z^{eq}_1 \times I^f_{A_1} - mz_1 \times I^f_{A_1}. \tag{6.24}$$

The negative-sequence voltage phasor at the fault point, $V^f_{F_2}$, can be expressed as:

$$V^f_{F_2} = -\left(Z^{eq}_1 + mz_1 \right) \times I^f_{A_2}. \tag{6.25}$$

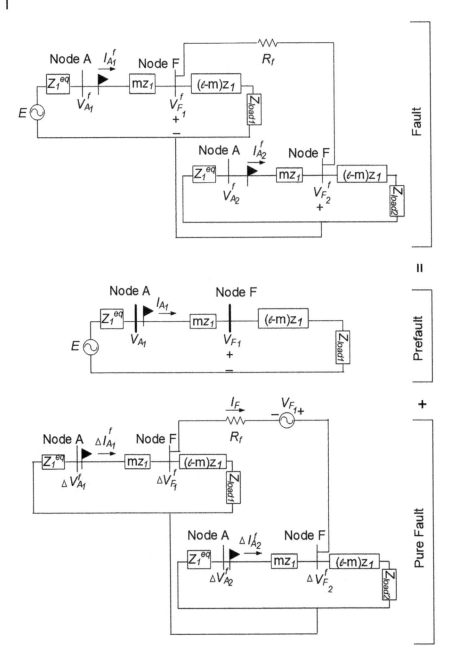

Figure 6.5 Superposition principle applied to the distribution network during a line-to-line fault. The flag indicates the location of the relay.

Substituting the expressions of $V^f_{F_1}$ and $V^f_{F_2}$ in (6.22) and assuming a bolted fault with zero fault resistance, the expression for fault distance can be written as

$$m = \frac{V_{A_1} - Z^{eq}_1\left(I^f_{A_1} - I^f_{A_2}\right)}{z_1\left(I^f_{A_1} - I^f_{A_2}\right)}. \tag{6.26}$$

Because $I^f_{A_1} = -I^f_{A_2}$, the above equation simplifies to:

$$m = \frac{V_{A_1} - 2 \times Z^{eq}_1 \times I^f_{A_1}}{2 \times z_1 \times I^f_{A_1}}. \tag{6.27}$$

Writing (6.27) in polar form,

$$m = \frac{|V_{A_1}|\angle\theta_{v_{a1}} - 2 \times |Z^{eq}_1| \times |I^f_{A_1}|\angle(\theta_z + \theta^f_{i_{a1}})}{2 \times |z_1| \times |I^f_{A_1}|\angle(\theta_{z_{l1}} + \theta^f_{i_{a1}})}, \tag{6.28}$$

where $\theta_{v_{a1}}$ and $\theta^f_{i_{a1}}$ are the phase angles of V_{A_1} and $I^f_{A_1}$, respectively. Since m is a real number, it can be estimated from only the magnitude terms as

$$m = \frac{\left||V_{A_1}|\angle\theta_{v_{a1}} - 2 \times |Z^{eq}_1| \times |I^f_{A_1}|\angle(\theta_z + \theta^f_{i_{a1}})\right|}{2 \times |z_1| \times |I^f_{A_1}|}. \tag{6.29}$$

Solving for m from the above equation is a challenging task as there is one equation but two unknowns: m and $\theta_{v_{a1}}$. To reduce the number of unknowns, we apply the reverse triangular inequality theorem on the numerator as:

$$m = \frac{\left||V_{A_1}| - 2 \times |Z^{eq}_1| \times |I^f_{A_1}|\right|}{2 \times |z_1| \times |I^f_{A_1}|} \quad \text{[miles]} \quad \because |u - v| \geq ||u| - |v||. \tag{6.30}$$

We assume that $\angle\theta_{v_{a1}}$ is equal to the summation of $\angle\theta_z$ and $\angle\theta^f_{i_{a1}}$. When this condition is not fulfilled, the numerator of (6.30) will be lower than its actual value, and the distance to fault location will be underestimated. Put another way, m can be regarded as the lower bound for the actual distance to the fault. Exercise 4 shows that the error due to this angle mismatch is not significant and (6.30) can accurately track down the location of a line-to-line fault.

(c) Single Line-to-Ground Fault

Suppose that a single line-to-ground fault with zero fault resistance occurs on phase A at a distance m from node A. The only information available to locate this fault are the current magnitudes of all the three phases at node A during the time of the fault. Neither the voltage sag of the faulted phase nor the phase angle relationship between the three currents are available.

To calculate fault location with the given data, the first step is to estimate the sequence current phasors at the relay location. Because $|I^f_{A_a}| \gg |I^f_{A_b}|$ and $|I^f_{A_c}|$ in an A-G fault, a simplifying assumption is made to ignore the load currents in the unfaulted phases. The sequence current calculation simplifies to:

$$
\begin{bmatrix} I^f_{A_0} \\ I^f_{A_1} \\ I^f_{A_2} \end{bmatrix} = \frac{1}{3} \begin{bmatrix} 1 & 1 & 1 \\ 1 & a & a^2 \\ 1 & a^2 & a \end{bmatrix} \times \begin{bmatrix} I^f_{A_a} \\ 0 \\ 0 \end{bmatrix}; \quad a = 1\angle 120° \tag{6.31}
$$

$$
\therefore I^f_{A_0} = I^f_{A_1} = I^f_{A_2} = \frac{I^f_{A_a}}{3} \quad [\text{A}].
$$

The next step is to derive an expression for V_{F_1}. This is achieved by applying Kirchhoff's voltage law and summing all the voltages around the pure fault loop shown in Figure 6.3:

$$
V_{F_1} = \left(Z_1^{eq} + mz_1 \right) \Delta I^f_{A_1} + \left(Z_1^{eq} + mz_1 \right) \Delta I^f_{A_2} + \left(Z_0^{eq} + mz_0 \right) \Delta I^f_{A_0}. \tag{6.32}
$$

The fault is assumed to have a zero fault resistance. Furthermore, because the system is assumed to be balanced before the fault, $\Delta I^f_{A_2} = I^f_{A_2}$ and $\Delta I^f_{A_0} = I^f_{A_0}$. The expression for the pure fault positive-sequence current $\Delta I^f_{A_1}$, on the other hand, is not as simple as its counterparts, as shown below:

$$
\Delta I^f_{A_1} = I^f_{A_1} - I_{A_1}. \tag{6.33}
$$

Because prefault current $I_{A_1} \ll I^f_{A_1}$, we approximate $\Delta I^f_{A_1}$ to be equal to $I^f_{A_1}$. The pure fault sequence current phasors are, therefore, equal to each other as shown below:

$$
\Delta I^f_{A_0} = \Delta I^f_{A_1} = \Delta I^f_{A_2} = I^f_{A_1}. \tag{6.34}
$$

Substituting (6.34) in (6.32) and rearranging the terms, the expression for calculating the distance to fault is:

$$
m = \frac{\dfrac{V_{F_1}}{I^f_{A_1}} - \left(2Z_1^{eq} + Z_0^{eq} \right)}{2z_1 + z_0}. \tag{6.35}
$$

We simplify the above expression by defining the following variables:

$$
2z_1 + z_0 = a + jb,
$$
$$
2Z_1^{eq} + Z_0^{eq} = c + jd.
$$

Substituting the variables into (6.35) results in:

$$
m = \frac{\dfrac{V_{F_1}}{I^f_{A_1}} - (c + jd)}{a + jb}. \tag{6.36}
$$

Writing (6.36) in polar form, the following is obtained:

$$
m = \frac{\dfrac{|V_{F_1}|}{|I^f_{A_1}|} \angle \left(\theta_{v_{f1}} - \theta^f_{i_{a1}} \right) - \sqrt{c^2 + d^2} \angle \tan^{-1} \left(\dfrac{d}{c} \right)}{\sqrt{a^2 + b^2} \angle \tan^{-1} \left(\dfrac{b}{a} \right)}, \tag{6.37}
$$

where $\theta_{v_{f1}}$ and $\theta_{i_{a1}}^f$ are the phase angles of V_{F_1} and $I_{A_1}^f$, respectively. Taking magnitude on both sides of (6.37):

$$m = \frac{\left| \dfrac{|V_{F_1}|}{|I_{A_1}^f|} \angle \left(\theta_{v_{f1}} - \theta_{i_{a1}}^f \right) - \sqrt{c^2 + d^2} \angle \tan^{-1} \left(\dfrac{d}{c} \right) \right|}{\sqrt{a^2 + b^2}}. \tag{6.38}$$

Applying the reverse triangular inequality theorem to the numerator of (6.38) and assuming $\angle \left(\theta_{v_{f1}} - \theta_{i_{a1}}^f \right) = \angle \tan^{-1} \left(\frac{d}{c} \right)$, we can solve for m as:

$$m = \frac{\left| \dfrac{|V_{F_1}|}{|I_{A_1}^f|} - \sqrt{c^2 + d^2} \right|}{\sqrt{a^2 + b^2}} \quad \text{[miles]}. \tag{6.39}$$

The equation above can locate single line-to-ground faults when the only information available is the current magnitude of the faulted phase. The calculation is easy to implement and requires the additional input of source and line impedance data. Voltage V_{F_1} is the magnitude of the prefault voltage phasor at the fault point and can be assumed to be 1 per unit. The assumption that $\angle \left(\theta_{v_{f1}} - \theta_{i_{a1}}^f \right) = \angle \tan^{-1} \left(\frac{d}{c} \right)$ may not always hold true. In such scenarios, (6.39) will underestimate the distance to the fault location and can be regarded as a lower bound for the actual distance to the fault.

6.3 Short-Circuit Fault Current Profile Method

This method uses fault current magnitude and circuit model of the distribution feeder to track down the fault location. The approach consists of simulating faults along the entire length of the feeder and building a reference current profile of short-circuit fault current versus distance to fault. In the event of a fault, the fault current magnitude is simply extrapolated on the current profile to get a location estimate. As an example, consider the utility circuit model shown in Figure 6.6. The model has been built using a power system modeling software and is considered to be an accurate representation of the actual feeder. You can see that the main trunk of the feeder has a couple of taps and three-phase underground loops. Building a short-circuit fault current profile for such a feeder is challenging since it is difficult to determine which feeder length should be chosen for building the current profile.

Reference [107] recommends building current profiles along every lateral and branch from the monitoring location. This strategy would result in the current profile having multiple subplots but would give accurate estimates since all the laterals and branches are taken into consideration. Adopting the same strategy for the feeder in our example, we identify four paths as illustrated in Figure 6.7. Path 1 is from the substation to the end of tap 1, path 2 is from the substation to the end of tap 2, path 3 is the main trunk and extends from the substation all the way up to the end of underground loop 1, and path 4 is from the substation to the end of underground loop 2. We stage single line-to-ground faults at successive incremental distances from the substation along a particular path. The corresponding fault current magnitude recorded at the substation is then plotted against the distance of the fault from the substation as shown in Figure 6.8. Because we have four paths, we have four

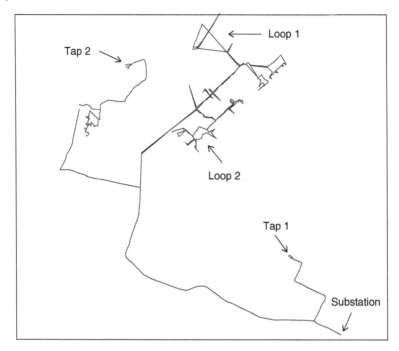

Figure 6.6 Circuit model of the distribution feeder in a power system modeling software.

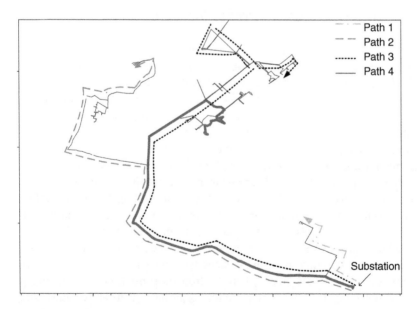

Figure 6.7 Paths for building the fault current profile.

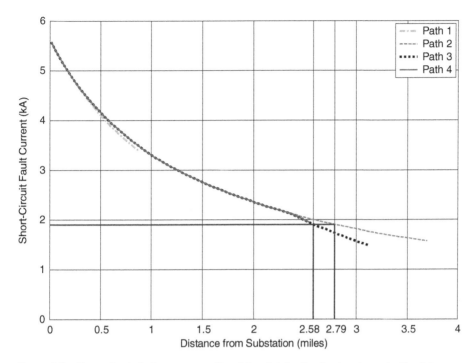

Figure 6.8 Short-circuit fault current profile of the distribution feeder shown in Fig. 6.6.

current profiles. Notice that the current profiles of all four paths are exponentially decaying curves. This is because short-circuit current decreases with distance from the substation. Now suppose that a single line-to-ground fault occurs on the feeder and that the relay at the substation records a fault current magnitude of 1900 A. To determine fault location, extrapolate 1900 A on the current profile as shown in Figure 6.8. The fault current intersects path 3 at 2.58 miles and path 2 at 2.79 miles. The fault could be at either of those two locations. Use some of the principles outlined in Chapter 5 to zero in on the best location estimate.

The method is highly accurate [108] since it makes no assumption about the distribution feeder being homogeneous. It does, however, require an accurate circuit model of the distribution feeder. It also assumes a bolted fault. As a result, locating faults that have a considerable fault resistance may be a challenging task.

6.4 Exercise Problems

■ Exercise 6.1

A distribution feeder experienced a BG fault. Line crew found a broken jumper 1.70 miles away from a digital relay protecting the feeder. Figure 6.9 shows a 15-cycle snapshot of the voltage and current seen by the relay before and during the fault. Notice how the B-phase fault current is not constant and decreases over time.

The relay in this exercise has measured the line-to-ground voltages during the fault. However, for the purpose of evaluating the effectiveness of the current-only algorithms, pretend

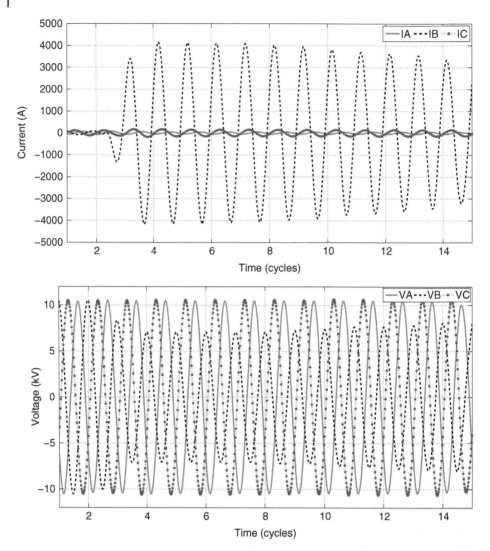

Figure 6.9 Exercise 6.1: Current and voltage waveforms captured by the digital relay during a B-G fault caused by a broken jumper.

that the voltage waveforms are not available for fault location. We will use the voltage measurements only when assessing the accuracy of the estimated fault voltage. The prefault $(I_{A_{abc}})$ current phasors, calculated by applying a fast Fourier filter on the first cycle, is given below. By looking at the phase angles, you can tell that the system has ABC phase rotation. The fault current phasors $(I^f_{A_{abc}})$ were calculated by applying the Fourier filter on the fourth cycle, which is when the fault was stable and had a higher current magnitude.

$$I_{A_{abc}} \text{ [A]} \qquad I^f_{A_{abc}} \text{ [A]}$$

$$\begin{bmatrix} 52.74\angle - 111.70° \\ 62.55\angle 122.93° \\ 82.11\angle - 7.00° \end{bmatrix} \qquad \begin{bmatrix} 36.31\angle - 128.29° \\ 2942.30\angle 24.42° \\ 115.13\angle - 17.18° \end{bmatrix}.$$

Setting	Description	Value
CTR	CT Ratio	120.00
PTR	PT Ratio	60.00
Z1MAG	Positive-Sequence Line Impedance Magnitude (ohms, sec)	6.30
Z1ANG	Positive-Sequence Line Impedance Angle (deg)	62.00
Z0MAG	Zero-Sequence Line Impedance Magnitude (ohms, sec)	16.80
Z0ANG	Zero-Sequence Line Impedance Angle (deg)	66.00
LL	Line Length	4.20
PHROT	Phase Rotation	ABC

Figure 6.10 Relay settings in Exercise 1.

The relay settings are given in Figure 6.10. Note that the line length is in units of miles. The rated voltage at the relay location is 12.47 kV. The positive- and zero-sequence source impedances behind the relay are $Z_1^{eq} = 0.0313 + j0.7602\,\Omega$ and $Z_0^{eq} = 0.0224 + j0.9624\,\Omega$, respectively. The distribution feeder is operating at 0.87 power factor (leading). Determine the following:

(a) Sequence fault current phasors measured by the relay.
(b) Sequence prefault current phasors measured by the relay.
(c) Positive-sequence prefault voltage phasor at the relay location.
(d) Positive-sequence pure fault voltage phasor at the relay location.
(e) Phase fault voltage phasors at the relay location.
(f) Sequence line impedances in primary ohms per mile.
(g) Zero-sequence compensation factor, k.
(h) Fault location.

Solution:

(a) Sequence fault current phasors, $I_{A_{012}}^f$, are obtained from the equation given below for ABC phase rotation. Because this is a BG fault, we use B-phase as the reference.

$$\begin{bmatrix} I_{A_0}^f \\ I_{A_1}^f \\ I_{A_2}^f \end{bmatrix} = \frac{1}{3} \begin{bmatrix} 1 & 1 & 1 \\ 1 & a & a^2 \\ 1 & a^2 & a \end{bmatrix} \times \begin{bmatrix} I_{A_b}^f \\ I_{A_c}^f \\ I_{A_a}^f \end{bmatrix}$$

$$= \frac{1}{3} \begin{bmatrix} 1 & 1 & 1 \\ 1 & a & a^2 \\ 1 & a^2 & a \end{bmatrix} \times \begin{bmatrix} 2942.30\angle24.42° \\ 115.13\angle-17.18° \\ 36.31\angle-128.29° \end{bmatrix}$$

$$= \begin{bmatrix} 999.18\angle22.64° \\ 990.29\angle27.30° \\ 954.70\angle23.30° \end{bmatrix} \quad [\text{A}].$$

(b) Sequence prefault current phasors, $I_{A_{012}}$, are obtained from the equation given below. Because we chose phase B as the reference when calculating the fault current phasors, we select phase B as the reference.

$$\begin{bmatrix} I_{A_0} \\ I_{A_1} \\ I_{A_2} \end{bmatrix} = \frac{1}{3} \begin{bmatrix} 1 & 1 & 1 \\ 1 & a & a^2 \\ 1 & a^2 & a \end{bmatrix} \times \begin{bmatrix} I_{A_b} \\ I_{A_c} \\ I_{A_a} \end{bmatrix}$$

$$= \frac{1}{3} \begin{bmatrix} 1 & 1 & 1 \\ 1 & a & a^2 \\ 1 & a^2 & a \end{bmatrix} \times \begin{bmatrix} 62.55\angle122.93° \\ 82.11\angle -7.00° \\ 52.74\angle -111.70° \end{bmatrix}$$

$$= \begin{bmatrix} 9.58\angle -13.09° \\ 65.38\angle120.23° \\ 10.57\angle -170.08° \end{bmatrix} \quad [\text{A}].$$

(c) The magnitude of the positive-sequence voltage phasor at the relay location just prior to the fault, V_{A_1}, can be assumed to be 1 per unit. In terms of volts primary, that becomes:

$$|V_{A_1}| = \frac{12.47 \times 1000}{\sqrt{3}} = 7199.60 \quad [\text{V}].$$

The phase angle of V_{A_1} can be calculated by rewriting (6.7) for a leading power factor condition as:

$$\theta_{v_{a1}} = \theta_{i_{a1}} - \cos^{-1}(\text{pf}) = 120.23° - \cos^{-1}(0.87) = 90.69°.$$

(d) The positive-sequence pure fault voltage phasor at the relay location, $\Delta V_{A_1}^f$, can be calculated using (6.8) as:

$$\Delta V_{A_1}^f = - \left[Z_1^{eq} \times \left(I_{A_1}^f - I_{A_1} \right) \right]$$

$$= -(0.0313 + j0.7602) \times \left(990.29\angle27.30° - 65.38\angle120.23° \right)$$

$$= 757.63\angle -68.82° \quad [\text{V}].$$

(e) The positive-sequence voltage phasor at the relay location can be calculated per (6.6) as:

$$V_{A_1}^f = V_{A_1} + \Delta V_{A_1}^f$$

$$= 7199.60\angle90.69° + 757.63\angle -68.82°$$

$$= 6495.30\angle88.35° \quad [\text{V}].$$

The negative-sequence voltage phasor at the relay location is given by (6.4) and can be calculated as shown below.

$$V_{A_2}^f = - \left(Z_1^{eq} \times I_{A_2}^f \right)$$

$$= -(0.0313 + j0.7602) \times 954.70\angle23.30°$$

$$= 726.38\angle -69.06° \quad [\text{V}].$$

The zero-sequence voltage phasor at the relay location is given by (6.5) and can be calculated as shown below.

$$V_{A_0}^f = - \left(Z_0^{eq} \times I_{A_0}^f \right)$$

$$= -(0.0224 + j0.9624) \times 999.18\angle22.64°$$

$$= 961.87\angle -68.69° \quad [\text{V}].$$

Using these sequence fault voltage phasors in (6.9), the phase voltages at the relay location during fault are obtained as:

$$
\begin{bmatrix} V^f_{A_b} \\ V^f_{A_c} \\ V^f_{A_a} \end{bmatrix} = \begin{bmatrix} 1 & 1 & 1 \\ 1 & a^2 & a \\ 1 & a & a^2 \end{bmatrix} \times \begin{bmatrix} V^f_{A_0} \\ V^f_{A_1} \\ V^f_{A_2} \end{bmatrix}
$$

$$
= \begin{bmatrix} 1 & 1 & 1 \\ 1 & a^2 & a \\ 1 & a & a^2 \end{bmatrix} \times \begin{bmatrix} 961.87\angle - 68.69° \\ 6495.30\angle 88.35° \\ 726.38\angle - 69.06° \end{bmatrix}
$$

$$
= \begin{bmatrix} 4982.10\angle 80.80° \\ 7358.00\angle - 30.55° \\ 7208.50\angle - 147.57° \end{bmatrix} \quad [V].
$$

You can see that the estimated fault voltage phasors compare quite well with the actual fault voltage phasors measured by the relay (shown below).

$$
\begin{bmatrix} V^f_{A_{b,act}} \\ V^f_{A_{c,act}} \\ V^f_{A_{a,act}} \end{bmatrix} = \begin{bmatrix} 5000.00\angle 90.00° \\ 7473.30\angle - 24.51° \\ 7441.90\angle - 141.82° \end{bmatrix} \quad [V].
$$

(f) The positive- and zero-sequence line impedance settings given in Figure 6.10 are in terms of ohms secondary. Because all our calculations are in terms of ohms primary, we use the equation below to convert the line impedance settings from ohms secondary to ohms primary [110].

$$
\Omega \text{ primary} = \Omega \text{ secondary} \times \frac{PTR}{CTR}. \tag{6.40}
$$

The positive- and zero-sequence line impedances in primary ohms can, therefore, be calculated as follows:

$$
Z_1 = 6.30\angle 62° \times \frac{60}{120} = 3.15\angle 62° \quad \Omega \text{ primary},
$$

$$
Z_0 = 16.80\angle 66° \times \frac{60}{120} = 8.40\angle 66° \quad \Omega \text{ primary}.
$$

We divide by line length to obtain the line impedances in primary ohms per mile.

$$
z_1 = \frac{Z_1}{LL} = \frac{3.15\angle 62°}{4.20} = 0.75\angle 62° \quad \Omega/\text{mile},
$$

$$
z_0 = \frac{Z_0}{LL} = \frac{8.40\angle 66°}{4.20} = 2.00\angle 66° \quad \Omega/\text{mile}.
$$

(g) The zero-sequence compensation factor, k can be determined as:

$$
k = \frac{z_0 - z_1}{z_1} = \frac{2.00\angle 66° - 0.75\angle 62°}{2.00\angle 66°} = 1.67\angle 6.39°.
$$

(h) Now that we have estimated the missing fault voltage phasors at the relay location, any one of the one-ended impedance-based methods can be used to calculate fault location. In this example, we make use of the simple reactance method. For a BG fault, the equation to calculate fault location is given below.

$$
m = \frac{\text{imag}\left(\dfrac{V^f_{A_b}}{I^f_{A_b} + k \times I^f_{A_0}}\right)}{\text{imag}(z_1)}
$$

$$
= \frac{\text{imag}\left(\dfrac{4982.10\angle 80.80°}{2942.30\angle 24.42° + 1.67\angle 6.39° \times 999.18\angle 22.64°}\right)}{\text{imag}(0.75\angle 62°)}
$$

$$
= 1.33 \quad [\text{mi}].
$$

The estimated location is close to the actual fault location of 1.70 miles.

■ **Exercise 6.2**

In the previous example, suppose that the only information available to you is a summary of the fault event shown in Figure 6.11. Waveform data is not available. From the event summary, we come to know that the relay saw a BG fault on 4/1/10 at 11:46 a.m. and measured a maximum current of 2979 A fault in the B phase. In addition, line impedance and source impedance behind the relay are also available. Determine fault location with the given data.

Solution:

We use (6.31) to determine the positive-sequence fault current magnitude as:

$$
|I^f_{A_1}| = \frac{I^f_{A_b}}{3} = \frac{2979}{3} = 993 \quad \text{A}.
$$

Using line and source impedance data (refer to the previous example), variables a, b, c, d evaluate to:

$$
a = \text{Re}\left(2z_1 + z_0\right) = 1.5177 \quad \Omega,
$$
$$
b = \text{Im}\left(2z_1 + z_0\right) = 3.1515 \quad \Omega,
$$
$$
c = \text{Re}\left(2Z^{eq}_1 + Z^{eq}_0\right) = 0.0850 \quad \Omega,
$$
$$
d = \text{Im}\left(2Z^{eq}_1 + Z^{eq}_0\right) = 2.4828 \quad \Omega.
$$

The prefault voltage at the fault point, V_{F_1}, is assumed to be 1 per unit, or 7199.60 V as shown by the calculation below.

$$
V_{F_1} = \frac{12{,}470}{\sqrt{3}} = 7199.60 \quad [\text{V}].
$$

#	DATE	TIME	EVENT	CURR	FREQ
1	04/01/10	11:46:44.118	BG	2979	60.01

Figure 6.11 Exercise 6.2: Event summary shows date and time of the fault, fault type, and maximum fault current magnitude. This is the only data available for fault location.

Substituting the above values in (6.39), the distance to fault from the relay can be calculated as:

$$m = \frac{\left| \dfrac{|V_{F_1}|}{|I^f_{A_1}|} - \sqrt{c^2 + d^2} \right|}{\sqrt{a^2 + b^2}}$$

$$= \frac{\left| \dfrac{7,199.60}{993} - \sqrt{0.0850^2 + 2.4828^2} \right|}{\sqrt{1.5177^2 + 3.1515^2}} = 1.36 \quad [\text{mi}].$$

The actual fault location is 1.70 miles. Given the limited amount of data available, the performance of the current magnitude–only method was excellent and close to the actual fault location.

■ Exercise 6.3

A tree fell on a distribution feeder at 0.95 miles from the substation and created a line-to-line fault between phases B and C. A digital relay at the substation captured the voltage and current waveforms before and during the fault and is shown in Figure 6.12.

For the purpose of proving the efficacy of the current-only fault location algorithms, pretend that the voltage waveforms are not available. The prefault ($I_{A_{abc}}$) and fault current phasors ($I^f_{A_{abc}}$), calculated by applying a fast Fourier filter on the first and the eighth cycle, respectively, are given below.

$$I_{A_{abc}}\,[\text{A}]$$
$$\begin{bmatrix} 157.33\angle - 167.33° \\ 167.38\angle 73.51° \\ 217.79\angle - 51.53° \end{bmatrix}$$

$$I^f_{A_{abc}}\,[\text{A}]$$
$$\begin{bmatrix} 180.29\angle - 164.23° \\ 2865.30\angle 64.08° \\ 2729.90\angle - 112.78° \end{bmatrix}.$$

Relay settings are given in Fig. 6.13. Note that the line length setting has been entered in units of miles. The distribution feeder is operating at 0.87 lagging power factor, and the positive-sequence source impedance behind the relay is known to be $Z^{eq}_1 = 0.1091 + j0.8633\,\Omega$. Determine the following:

(a) Sequence fault current phasors measured by the relay.
(b) Sequence prefault current phasors measured by the relay.
(c) Positive-sequence prefault voltage phasor at the relay location.
(d) Positive-sequence pure fault voltage phasor at the relay location.
(e) Phase fault voltage phasors at the relay location.
(f) Positive-sequence line impedance in primary ohms per mile.
(g) Fault location.

Solution:

(a) Sequence fault current phasors, $I^f_{A_{012}}$, are obtained from the equation given below for ABC phase rotation. Because this is a BC fault, we use A-phase as the reference.

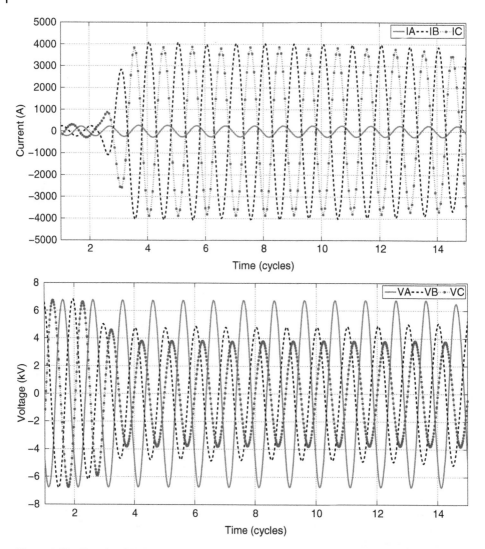

Figure 6.12 Exercise 6.3: Current and voltage waveforms captured by a digital relay at the substation during a line-to-line fault between phases B and C. The two faulted phases are equal in magnitude but 180 degrees out of phase with each other.

Setting	Description	Value
CTR	CT Ratio	240.00
PTR	PT Ratio	40.00
VNOM	Phase Nominal Voltage (V, sec)	120.00
Z1MAG	Positive-Sequence Line Impedance Magnitude (ohms, sec)	9.00
Z1ANG	Positive-Sequence Line Impedance Angle (deg)	54.00
LL	Line Length	2.42
PHROT	Phase Rotation	ABC

Figure 6.13 Exercise 6.3: Relay settings.

$$
\begin{bmatrix} I^f_{A_0} \\ I^f_{A_1} \\ I^f_{A_2} \end{bmatrix} = \frac{1}{3} \begin{bmatrix} 1 & 1 & 1 \\ 1 & a & a^2 \\ 1 & a^2 & a \end{bmatrix} \times \begin{bmatrix} I^f_{A_a} \\ I^f_{A_b} \\ I^f_{A_c} \end{bmatrix}
$$

$$
= \frac{1}{3} \begin{bmatrix} 1 & 1 & 1 \\ 1 & a & a^2 \\ 1 & a^2 & a \end{bmatrix} \times \begin{bmatrix} 180.29\angle -164.23° \\ 2865.30\angle 64.08° \\ 2729.90\angle -112.78° \end{bmatrix}
$$

$$
= \begin{bmatrix} 8.20\angle 26.72° \\ 1687.20\angle 157.70° \\ 1544.30\angle -26.66° \end{bmatrix} \quad [A].
$$

(b) Sequence prefault current phasors, $I_{A_{012}}$, are obtained from the equation given below. We select phase A as the reference since this phase was the reference when calculating the sequence current phasors during the fault.

$$
\begin{bmatrix} I_{A_0} \\ I_{A_1} \\ I_{A_2} \end{bmatrix} = \frac{1}{3} \begin{bmatrix} 1 & 1 & 1 \\ 1 & a & a^2 \\ 1 & a^2 & a \end{bmatrix} \times \begin{bmatrix} I_{A_a} \\ I_{A_b} \\ I_{A_c} \end{bmatrix}
$$

$$
= \frac{1}{3} \begin{bmatrix} 1 & 1 & 1 \\ 1 & a & a^2 \\ 1 & a^2 & a \end{bmatrix} \times \begin{bmatrix} 157.33\angle -167.33° \\ 167.38\angle 73.51° \\ 217.79\angle -51.53° \end{bmatrix}
$$

$$
= \begin{bmatrix} 17.80\angle -56.48° \\ 180.69\angle -168.76° \\ 20.86\angle 48.25° \end{bmatrix} \quad [A].
$$

(c) The magnitude of the positive-sequence voltage phasor at the relay location just prior to the fault, V_{A_1}, can be assumed to be 1 per unit. In terms of volts primary, that becomes:

$$
|V_{A_1}| = \text{VNOM} \times \text{PTR} = 120 \times 40 = 4800 \quad [V].
$$

The phase angle of V_{A_1} can be calculated using (6.7) as:

$$
\theta_{v_{a1}} = \cos^{-1}(\text{pf}) + \theta_{i_{a1}} = \cos^{-1}(0.87) - 168.76° = -139.22°.
$$

(d) The positive-sequence pure fault voltage phasor at the relay location, $\Delta V^f_{A_1}$, can be calculated using (6.8) as:

$$
\Delta V^f_{A_1} = -\left[Z^{eq}_1 \times \left(I^f_{A_1} - I_{A_1} \right) \right]
$$

$$
= -\left[(0.1091 + j0.8633) \times (1687.20\angle 157.70° - 180.69\angle -168.76°) \right]
$$

$$
= 1339.91\angle 56.78° \quad [V].
$$

(e) Let us start by first determining the positive-sequence voltage phasor at the relay location using (6.6) as:

$$
V^f_{A_1} = V_{A_1} + \Delta V^f_{A_1}
$$

$$
= 4800\angle -139.22° + 1339.91\angle 56.78°
$$

$$
= 3531.36\angle -145.22° \quad [V].
$$

The negative-sequence voltage phasor at the relay location is given by (6.4) and can be calculated as shown below.

$$V^f_{A_2} = -\left(Z^{eq}_1 \times I^f_{A_2}\right)$$

$$= -(0.1091 + j0.8633) \times 1544.30\angle -26.66°$$

$$= 1343.80\angle -123.86° \quad [V].$$

Because zero-sequence quantities are negligible during a line-to-line fault, set the zero-sequence voltage phasor at the relay location, $V^f_{A_0}$, equal to zero. Using these sequence fault voltage phasors in (6.9), the phase voltages at the relay location during fault are obtained as:

$$
\begin{bmatrix} V^f_{A_a} \\ V^f_{A_b} \\ V^f_{A_c} \end{bmatrix}
=
\begin{bmatrix} 1 & 1 & 1 \\ 1 & a^2 & a \\ 1 & a & a^2 \end{bmatrix}
\times
\begin{bmatrix} V^f_{A_0} \\ V^f_{A_1} \\ V^f_{A_2} \end{bmatrix}
$$

$$
=
\begin{bmatrix} 1 & 1 & 1 \\ 1 & a^2 & a \\ 1 & a & a^2 \end{bmatrix}
\times
\begin{bmatrix} 0 \\ 3531.36\angle -145.22° \\ 1343.80\angle -123.86° \end{bmatrix}
$$

$$
=
\begin{bmatrix} 4807.90\angle -139.38° \\ 3584.60\angle 73.03° \\ 2619.80\angle -6.54° \end{bmatrix} \quad [V].
$$

You can see that the estimated fault voltage phasors compare quite well with the actual fault voltage phasors measured by the relay in this example (shown below).

$$
\begin{bmatrix} V^f_{A_a,act} \\ V^f_{A_b,act} \\ V^f_{A_c,act} \end{bmatrix}
=
\begin{bmatrix} 4775.90\angle -135.42° \\ 3401.50\angle 77.63° \\ 2668.10\angle 0.58° \end{bmatrix} \quad [V].
$$

(f) The positive-sequence line impedance setting given in Fig. 6.13 is in terms of ohms secondary. We convert the impedance setting from ohms secondary to ohms primary using the equation below.

$$\Omega \text{ primary} = \Omega \text{ secondary} \times \frac{\text{PTR}}{\text{CTR}} = 9\angle 54° \times \frac{40}{240} = 1.5\angle 54°.$$

Next, we obtain the line impedance in terms of primary ohms per mile as:

$$z_1 = \frac{\Omega \text{ primary}}{\text{LL}} = \frac{1.5\angle 54°}{2.42} = 0.6198\angle 54° \quad \Omega/\text{mile}.$$

(g) Now that we have estimated the missing fault voltage phasors at the relay location, any one of the one-ended impedance-based methods can be used to calculate fault location. In this example, we make use of the simple reactance method. Because this is a BC fault, the equation to calculate fault location is given below.

$$m = \frac{\text{imag}\left(\dfrac{V^f_{A_b} - V^f_{A_c}}{I^f_{A_b} - I^f_{A_c}}\right)}{\text{imag}(z_1)}$$

$$= \frac{\text{imag}\left(\dfrac{3584.60\angle 73.03° - 2619.80\angle -6.54°}{2865.30\angle 64.08° - 2729.90\angle -112.78°}\right)}{\text{imag}(0.6198\angle 54°)}$$

$$= 1.05 \quad [\text{mi}].$$

The actual fault location is known to be 0.95 miles. The example illustrates that the current-only fault location algorithm can successfully locate faults with only current phasors.

■ Exercise 6.4

In the previous example, suppose that the relay technician does not have a computer and is, hence, not able to download the waveforms captured by the digital relay. He accesses the event summary though the relay front panel and notes down the maximum current magnitudes seen by the relay during the fault: 196 A on phase A, 2910 A on phase B, and 2750 A on phase C. However, because the actual waveform data could not be downloaded, the phase angle relation between the three phase currents is not known. The event summary also indicates that the relay saw a BC fault. Calculate fault location using only the fault current magnitudes.

Solution:

We begin by reconstructing the fault current phasors as shown below.

$$I^f_{A_a} = 0 \quad [A],$$

$$I^f_{A_b} = 2910\angle 50° \quad [A],$$

$$I^f_{A_c} = -I^f_{A_b} \quad [A].$$

We then proceed to estimate the sequence current phasors by using phase A as the reference as shown below.

$$\begin{bmatrix} I^f_{A_0} \\ I^f_{A_1} \\ I^f_{A_2} \end{bmatrix} = \frac{1}{3}\begin{bmatrix} 1 & 1 & 1 \\ 1 & a & a^2 \\ 1 & a^2 & a \end{bmatrix} \times \begin{bmatrix} I^f_{A_a} \\ I^f_{A_b} \\ I^f_{A_c} \end{bmatrix}; \quad a = 1\angle 120°$$

$$= \frac{1}{3}\begin{bmatrix} 1 & 1 & 1 \\ 1 & a & a^2 \\ 1 & a^2 & a \end{bmatrix} \times \begin{bmatrix} 0 \\ 2910\angle 50° \\ 2910\angle -130° \end{bmatrix},$$

$$\therefore I^f_{A_0} = 0; \quad I^f_{A_1} = -I^f_{A_2} = 1680.10\angle 140° \quad [A].$$

The magnitude of the positive-sequence voltage phasor at the relay location just prior to the fault, V_{A_1}, can be assumed to be 1 per unit, or 4800 V (see previous example). We can now calculate the distance to fault using (6.30) as:

$$m = \frac{\left| |V_{A_1}| - 2 \times |Z^{eq}_1| \times |I^f_{A_1}| \right|}{2 \times |z_1| \times |I^f_{A_1}|}$$

$$= \frac{|4800 - (2 \times 0.8702 \times 1680.10)|}{2 \times 0.6198 \times 1680.10}$$

$$= 0.90 \quad [mi].$$

The actual fault location is known to be 0.95 miles.

■ Exercise 6.5

A 24.9 kV distribution feeder experienced a three-phase fault when strong winds caused a tree to fall, breaking a cross arm of the distribution pole and creating a fault. Line crew found the tree and the damaged cross arm at a distance of 4.27 miles from the substation. A digital

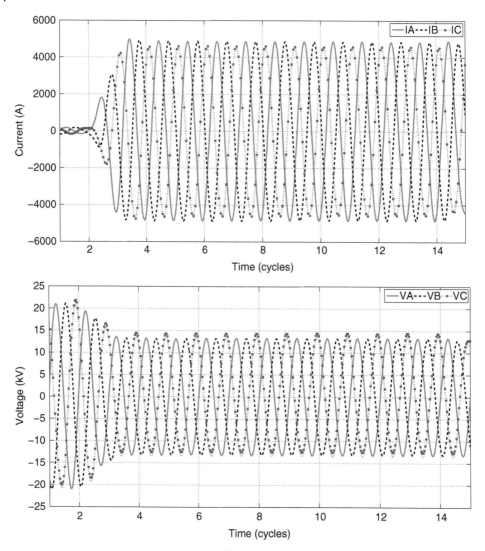

Figure 6.14 Exercise 6.5: Current and voltage data recorded by a digital relay at the substation during a three-phase fault.

relay at the substation had captured both the voltage and current waveforms during the fault. The waveforms are shown in Figure 6.14. Pretend that the voltage waveforms are not available for calculating fault location. We do this to demonstrate the efficacy of the current-only fault location algorithms. The prefault and fault current phasors, calculated by applying a fast Fourier filter to the first and eleventh cycle, respectively, are given below.

$$
\begin{array}{cc}
I_{A_{abc}}[A] & I^f_{A_{abc}}[A] \\
\begin{bmatrix} 120.62\angle 11.96° \\ 131.92\angle -109.95° \\ 123.07\angle 128.73° \end{bmatrix} &
\begin{bmatrix} 3440.00\angle -67.08° \\ 3437.90\angle 169.79° \\ 3267.60\angle 52.06° \end{bmatrix}.
\end{array}
$$

Setting	Description	Value
CTR	CT Ratio	240.00
PTR	PT Ratio	120.00
VNOM	Phase Nominal Voltage (V, sec)	120.00
Z1MAG	Positive-Sequence Line Impedance Magnitude (ohms, sec)	12.40
Z1ANG	Positive-Sequence Line Impedance Angle (deg)	66.00
LL	Line Length	9.14
PHROT	Phase Rotation	ABC

Figure 6.15 Exercise 6.5: Relay settings.

The relay settings are given in Figure 6.15. The positive-sequence source impedance behind the relay is known to be $Z_1^{eq} = 0.0604 + j1.6568\,\Omega$. The distribution feeder is operating at almost unity power factor. Determine the following:

(a) Sequence fault current phasors measured by the relay.
(b) Sequence prefault current phasors measured by the relay.
(c) Positive-sequence prefault voltage phasor at the relay location.
(d) Positive-sequence pure fault voltage phasor at the relay location.
(e) Phase fault voltage phasors at the relay location.
(f) Positive-sequence line impedance in primary ohms per mile.
(g) Fault location.

Solution:

(a) Sequence fault current phasors, $I_{A_{012}}^f$, are obtained from the equation given below for ABC phase rotation. Choose A-phase as the reference.

$$\begin{bmatrix} I_{A_0}^f \\ I_{A_1}^f \\ I_{A_2}^f \end{bmatrix} = \frac{1}{3}\begin{bmatrix} 1 & 1 & 1 \\ 1 & a & a^2 \\ 1 & a^2 & a \end{bmatrix} \times \begin{bmatrix} I_{A_a}^f \\ I_{A_b}^f \\ I_{A_c}^f \end{bmatrix}$$

$$= \frac{1}{3}\begin{bmatrix} 1 & 1 & 1 \\ 1 & a & a^2 \\ 1 & a^2 & a \end{bmatrix} \times \begin{bmatrix} 3440.00\angle -67.08° \\ 3437.90\angle 169.79° \\ 3267.60\angle 52.06° \end{bmatrix}$$

$$= \begin{bmatrix} 13.10\angle 152.69° \\ 3380.90\angle -68.42° \\ 111.90\angle -15.83° \end{bmatrix} \quad \text{[A]}.$$

(b) Sequence prefault current phasors, $I_{A_{012}}$, are obtained from the equation given below using A-phase as the reference (same phase as the one used as the reference for calculating sequence current phasors during the fault).

$$
\begin{bmatrix} I_{A_0} \\ I_{A_1} \\ I_{A_2} \end{bmatrix} = \frac{1}{3} \begin{bmatrix} 1 & 1 & 1 \\ 1 & a & a^2 \\ 1 & a^2 & a \end{bmatrix} \times \begin{bmatrix} I_{A_a} \\ I_{A_b} \\ I_{A_c} \end{bmatrix}
$$

$$
= \frac{1}{3} \begin{bmatrix} 1 & 1 & 1 \\ 1 & a & a^2 \\ 1 & a^2 & a \end{bmatrix} \times \begin{bmatrix} 120.62\angle 11.96° \\ 131.92\angle -109.95° \\ 123.07\angle 128.73° \end{bmatrix}
$$

$$
= \begin{bmatrix} 1.67\angle -143.27° \\ 125.17\angle 10.23° \\ 5.37\angle 135.60° \end{bmatrix} \quad [\text{A}].
$$

(c) The magnitude of the positive-sequence voltage phasor at the relay location just prior to the fault, V_{A_1}, can be assumed to be 1 per unit. In terms of volts primary, that becomes:

$$
|V_{A_1}| = \text{VNOM} \times \text{PTR} = 120 \times 120 = 14,400 \quad [\text{V}].
$$

Because the distribution feeder is operating at unity power factor, set the phase angle of V_{A_1} equal to that of I_{A_1}.

(d) The positive-sequence pure fault voltage phasor at the relay location, $\Delta V_{A_1}^f$, can be calculated using (6.8) as:

$$
\Delta V_{A_1}^f = -\left[Z_1^{eq} \times \left(I_{A_1}^f - I_{A_1} \right) \right]
$$

$$
= -\left[(0.0604 + j1.6568) \times \left(3380.90\angle -68.42° - 125.17\angle 10.23° \right) \right]
$$

$$
= 5568.07\angle -162.60° \quad [\text{V}].
$$

(e) Let us start by first determining the positive-sequence voltage phasor at the relay location using (6.6) as:

$$
V_{A_1}^f = V_{A_1} + \Delta V_{A_1}^f
$$

$$
= 14,400\angle 10.23° + 5568.07\angle -162.60°
$$

$$
= 8902.60\angle 5.75° \quad [\text{V}].
$$

Because this is a balanced fault, we can set the negative- and zero-sequence voltage phasors equal to zero. Substitute the sequence fault voltage phasors in (6.9) to obtain the phase voltages at the relay location during fault as:

$$
\begin{bmatrix} V_{A_a}^f \\ V_{A_b}^f \\ V_{A_c}^f \end{bmatrix} = \begin{bmatrix} 1 & 1 & 1 \\ 1 & a^2 & a \\ 1 & a & a^2 \end{bmatrix} \times \begin{bmatrix} V_{A_0}^f \\ V_{A_1}^f \\ V_{A_2}^f \end{bmatrix}
$$

$$
= \begin{bmatrix} 1 & 1 & 1 \\ 1 & a^2 & a \\ 1 & a & a^2 \end{bmatrix} \times \begin{bmatrix} 0 \\ 8902.60\angle 5.75° \\ 0 \end{bmatrix}
$$

$$
= \begin{bmatrix} 8902.60\angle 5.75° \\ 8902.60\angle -114.25° \\ 8902.60\angle 125.75° \end{bmatrix} \quad [\text{V}].
$$

(f) The positive-sequence line impedance setting given in Figure 6.15 is in terms of ohms secondary. We convert the impedance setting from ohms secondary to ohms primary using the equation below.

$$\Omega \text{ primary} = \Omega \text{ secondary} \times \frac{\text{PTR}}{\text{CTR}} = 12.40\angle66° \times \frac{120}{240} = 6.2\angle66°.$$

Next, we obtain the line impedance in terms of primary ohms per mile as:

$$z_1 = \frac{\Omega \text{ primary}}{\text{LL}} = \frac{6.2\angle66°}{9.14} = 0.6783\angle66° \quad \Omega/\text{mile}.$$

(g) Now that we have estimated the missing fault voltage phasors at the relay location, any one of the one-ended impedance-based methods can be used to calculate fault location. In this example, we make use of the simple reactance method. The equation to calculate fault location is given below.

$$
\begin{aligned}
m &= \frac{\text{imag}\left(\dfrac{V_{A_b}^f - V_{A_c}^f}{I_{A_b}^f - I_{A_c}^f}\right)}{\text{imag}(z_1)} \\[2mm]
&= \frac{\text{imag}\left(\dfrac{8902.60\angle-114.25° - 8902.60\angle125.75°}{3437.90\angle169.79° - 3267.60\angle52.06°}\right)}{\text{imag}(0.6783\angle66°)} \\[2mm]
&= 4.20 \quad [\text{mi}].
\end{aligned}
$$

The actual fault location is known to be 4.27 miles.

■ Exercise 6.6

In the previous example, suppose that field personnel do not have access to the fault waveform data. The only data available from the relay front panel are the maximum current magnitudes seen by the relay: 3563 A in phase A, 3395 A in phase B, and 3290 A in phase C. Determine fault location using fault current magnitude only.

Solution:

We perform all our calculations in this example with C-phase. The C-phase voltage magnitude prior to the fault, $|V_{A_c}|$, is 1 per unit, or 14,400 V (see previous example). The C-phase voltage magnitude during fault can be calculated using (6.15) as:

$$
\begin{aligned}
\left|V_{A_c}^f\right| &= \left||V_{A_c}| - \left(|Z_1^{eq}| \times |I_{A_c}^f|\right)\right| \\
&= |14400 - (1.6579 \times 3290)| \\
&= 8945.51 \quad \text{V}.
\end{aligned}
$$

The distance to fault is then calculated using (6.19) as

$$
\begin{aligned}
m &= \frac{|V_{A_c}^f|}{|z_1| \times |I_{A_c}^f|} \\
&= \frac{8945.51}{0.6783 \times 3290} = 4.01 \text{ [mi]}
\end{aligned}
$$

The actual fault was found to be at 4.27 miles. Because all three phases have slightly different fault current magnitudes, you could also calculate fault location using each phase (not just the C phase) and then take the average of the three results.

6.5 Summary

This chapter proposes practical fault location algorithms that use current data only to estimate the distance to a fault. Two fault-locating approaches using current phasors and current magnitude are developed. Both approaches work by estimating fault voltage at the monitoring location and then invoking impedance-based fault-locating principles. Because of more data availability, the algorithm using current phasors makes no assumptions when estimating the distance to a fault and hence has a superior performance. It is also capable of locating all fault types. The algorithm using current magnitude, on the other hand, estimates fault location with limited data and can locate single line-to-ground, line-to-line, and three-phase faults. It also makes several assumptions such as an unloaded distribution feeder and bolted fault when calculating the fault location. Example exercises in this chapter show that in spite of all the assumptions, the algorithm is powerful and can track down the exact location of a fault. The chapter also presents another approach that uses the fault current magnitude and the system model for fault location purposes. This approach, referred to as the short-circuit fault current profile, can successfully locate any fault type.

7

System and Operational Benefits of Fault Location

The previous chapters discussed how to locate faults on transmission lines and distribution feeders. Knowledge of the fault location has the obvious benefit of reducing outage times but also has several system and operational benefits. First, fault location is a valuable input when analyzing relay operations. It allows protection engineers to determine whether the protection system operated as expected for a given fault. Second, investigating the why behind incorrect fault location results from a relay can reveal wiring errors or erroneous line impedance settings, both of which can lead to relay misoperations. Third, fault location can be used to estimate the fault resistance. Knowing the fault resistance value is important when analyzing the operation of a distance relay. The fault resistance value is also useful in gaining insight into the root cause of the fault. For example, fault impedance statistics developed by analyzing 148 fault events collected from several utility circuits suggest that trees with a larger diameter present a fault resistance greater than 20 ohms when they fall on overhead lines [115]. Fourth, fault location along with the fault resistance value can help validate the system short-circuit model. This can be achieved by replicating the same fault in the system model and comparing the output of the model against field measurements. A close match indicates that the model is representative of the actual system. Evaluating the accuracy of the short-circuit model is important as relay settings and circuit breaker ratings are determined with the help of this model. Fifth, fault location can be a valuable input to a protective relay that initiates autoreclosing in hybrid lines. A hybrid line consists of both underground cable and overhead line sections. If the fault is on the overhead section, it is desired that autoreclosing proceed since overhead faults are usually temporary faults. However, if the fault is on the underground cable section, it is desired that autoreclosing be blocked to prevent any further damage since cable faults are usually permanent faults. Fault location can help the reclosing relay determine whether the fault is on the overhead or underground section and whether it should reclose or not. Sixth, fault location can help utilities determine whether there is a relay coordination issue or whether multiple faults were involved in a trip-reclose-lockout sequence. Finally, tracking the location of temporary faults over a period of time can help identify weak spots on the power system and allow an opportunity for corrective action to be taken before these faults evolve into a permanent fault and cause an outage. This chapter explores the additional benefits of fault location in more detail.

Fault Location on Transmission and Distribution Lines: Principles and Applications, First Edition.
Swagata Das, Surya Santoso, and Sundaravaradan N. Ananthan.
© 2022 John Wiley & Sons Ltd. Published 2022 by John Wiley & Sons Ltd.
Companion website: www.wiley.com/go/das/faultlocation

7.1 Verify Relay Operation

When analyzing the operation of a protective relay, the first task is to know the location of the fault. This information plays a crucial role in determining whether the fault was within the zone of the protective relay and whether the protective relay operation was correct. Now is a good time to talk about zones of protection.

Figure 7.1 shows a simple schematic of the electrical power system with generators, step-up transformers, transmission lines, step-down transformers, distribution feeders, and loads. Since the different pieces of the power system are at different voltage levels and are geographically far apart, protecting this system with one protective relay is an impossible task. Furthermore, it is important that the protection scheme be selective and isolate only that portion of the power system that has experienced a fault. As an example, a fault on the distribution feeder should be cleared by the relay protecting that specific feeder. It should not cause a relay protecting the distribution bus to operate and result in a power outage for customers on the healthy feeders.

The solution is to divide up the power system into different zones and assign protective relays to be responsible for a particular zone as shown in Figure 7.1. The primary relays also act as a backup to protective relays in neighboring zones [38]. The zone is built around the power system equipment being protected and is defined by the location of the CTs and PTs. A zone should also have a circuit breaker or a device capable of isolating the fault within the zone of protection. Notice that zones of protection always overlap with each other. This

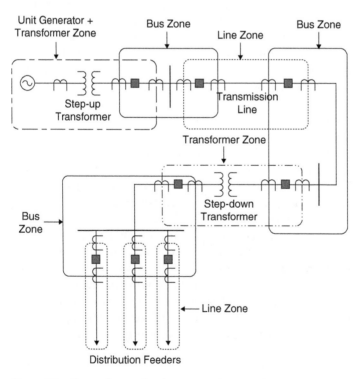

Figure 7.1 Zones of protection in an example power system.

is done to avoid any blind spot and to ensure that every part of the power system is protected. Typically the overlap is achieved by using CTs on either side of the circuit breaker. When there is a fault in the overlap area, primary relays of both zones will trip. By using the circuit breaker to overlap, the area of overlap is minimized. This reduces the probability of a fault occurring in the overlap area and the chances for both zones tripping for the same fault.

From the above discussion, it is evident that when relay engineers begin their analysis after a power system fault, they need to know, first and foremost, the location of the fault. Second, they need necessary documentation that explains the protection scheme and philosophy. Both help the engineer understand the expected operation during a power system fault. Comparing the expected with the actual operation helps determine whether the protection system operated as expected.

7.2 Discover Erroneous Relay Settings

Fault location analysis can reveal errors with line impedance settings that would have otherwise gone undetected. This is a significant benefit as a NERC report [111] published in 2013 concluded that incorrect relay settings, logic, or design errors were responsible for 28 percent of the misoperations between January 1, 2011, and April 1, 2012, as shown in Figure 7.2. Line impedance settings play a critical role in the correct operation of distance relays. Typically, zone 1 is set to reach up to 80 percent of the line impedance in the forward

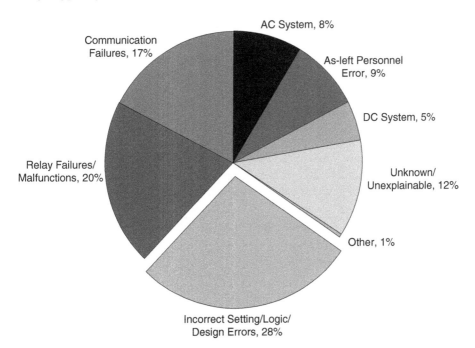

Figure 7.2 Misoperations between January 1, 2011, and April 1, 2012, categorized by root cause in a report published by NERC [111].

direction while zone 2 is set to overreach and covers 120 percent of the line in the forward direction. During a fault, if the relay measures an apparent impedance lower than the zone reach setting, it calls upon the circuit breaker to open and clear the fault from the transmission network [55]. A zone 1 fault will be cleared instantaneously. A zone 2 fault will be cleared after an intentional time delay. As you can see, inaccurate line impedance settings will affect the relay's decision to distinguish between an in-zone and out-of-zone fault and can result in a misoperation. In addition to distance relays, some directional relays also use line impedance data to determine whether the fault is in the forward or in the reverse direction. They will also be affected by incorrect line impedance settings. Therefore, if fault location analysis can help identify setting errors, it will certainly help reduce the number of misoperations. This section illustrates this with an example in which an incorrect fault location estimate, although frustrating, proved to be a blessing in disguise as it led to the discovery of wrong line impedance settings.

A distribution feeder serving 743 customers experienced a series of faults starting at 21:18 hours on July 22, 2010. A microprocessor-based digital relay at the 24.9 kV substation reclosed twice to allow the fault to clear away on its own. The reclose attempts, however, failed to clear the fault, and the relay locked out, causing all 743 customers to experience a sustained interruption for 176 minutes. The relay at the substation recorded five event reports during the duration of the fault. Figure 7.3 shows a summary of all the five event reports. Event 1 is the oldest event and event 5 is the latest event. Figure 7.4 shows the voltage and current waveform captured in each report.

Let's start with event 1 and reconstruct the sequence of events. During this process, we are going to refer to both Figure 7.3 and Figure 7.4. In event 1, the relay detected a line-to-line fault between phases B and C as indicated by the waveforms in Figure 7.4(a) and issued a trip to the circuit breaker at the substation (T in Fig. 7.3 stands for trip). The maximum phase current measured during the fault was 2600 A. This relay, in addition to protecting the distribution feeder, serves as a fault locator and uses the modified Takagi method to estimate the distance to a fault. The relay estimated the BC fault to have occurred at a distance of 5.46 miles from the substation. After allowing the fault sufficient time to clear out on its own (open interval time set by the utility), the relay issued a close to the circuit breaker at the substation. The fault was still present but manifested itself as a B-G fault as indicated by the waveforms in Figure 7.4(b) at an estimated location of 4.48 miles from the substation (event 2). The maximum phase current measured on the B-phase was 2210 A. The phase and ground time overcurrent elements in the relay picked up and started to time.

EVENT	DATE	TIME STAMP	FAULT TYPE	FAULT LOCATION	SHOT	MAX PHASE CURRENT
1	7/22/10	21:18:58.629	BC T*	5.46	0	2600
2	7/22/10	21:19:02.000	BG	4.48	1	2210
3	7/22/10	21:19:02.945	BC T*	5.22	1	2670
4	7/22/10	21:19:31.458	BG	4.51	2	2190
5	7/22/10	21:19:32.495	BCG T*	5.34	2	2820

* T stands for trip

Figure 7.3 Event history saved in the substation relay on the day of the fault.

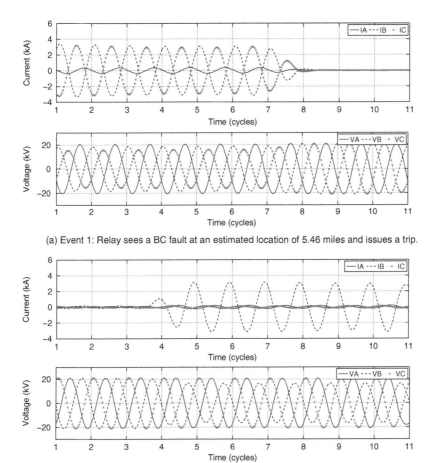

(a) Event 1: Relay sees a BC fault at an estimated location of 5.46 miles and issues a trip.

(b) Event 2: On first reclose, relay sees a B-G fault at an estimated location of 4.48 miles. The phase and ground overcurrent elements pick up and start to time.

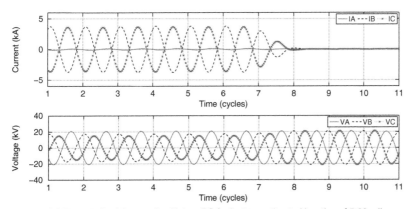

(c) Event 3: Fault has evolved into a BC fault at an estimated location of 5.22 miles. The relay issues a trip.

Figure 7.4 Waveforms captured by the digital relay during a permanent fault. The relay reclosed twice but was unable to clear the fault. It eventually locked out.

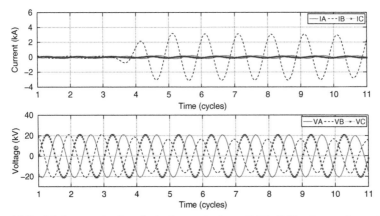

(d) Event 4: On second reclose, the fault comes back as a B-G fault. The phase and ground time overcurrent elements in the relay pick up and start to time. Estimated location is 4.51 miles.

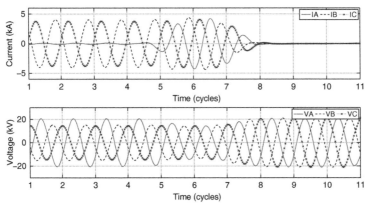

(e) Event 5: The fault evolves to a BC-G and finally to a three-phase fault. Relay issues a trip and locks out. Estimated location is 5.34 miles.

Figure 7.4 (Continued)

By the next event, event 3, the fault had evolved into a line-to-line fault between phase B and phase C. The phase time overcurrent element timed out and the relay issued a trip. The maximum phase current measured by the relay was 2670 A. The fault was estimated to be 5.22 miles from the substation. After letting the circuit breaker stay open for a settable time programmed by the utility, the relay issued a close. Unfortunately, the fault reappeared as a B-G fault (event 4) at an estimated location of 4.51 miles. The maximum phase current was measured to be 2190 A. The phase and ground time overcurrent elements picked up and started to time on their respective curves. In the final event (event 5), the fault had evolved into a BC-G fault and then into a three-phase fault. When the phase time overcurrent element timed out, the relay issued a trip and locked out. Estimated location of the fault was 5.34 miles. The maximum current seen by the relay in this event was 2820 A.

Notice how the fault location changes with the fault type in Figure 7.3? At first glance, it appears that the distribution feeder experienced two faults at two separate locations, a B-G fault between 4.48 and 4.51 miles and a BC fault that evolved into a BC-G fault between

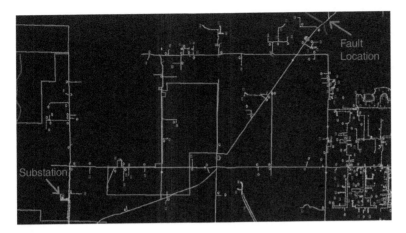

Figure 7.5 Utility network diagram showing the fault location.

5.22 and 5.46 miles. During line inspection, however, maintenance crew found evidence of just one fault. A jumper cable was found to be burned open on a distribution pole at 4.43 miles from the substation as shown in Figure 7.5. The distance of 4.43 miles was, therefore, reported as the actual fault location. After replacing the jumper, the distribution feeder was reenergized by manually closing the breaker at the substation.

From the above discussion, it is evident that the relay located the B-G fault correctly. However, it experienced an error of about a mile when trying to locate the BC or BC-G fault. We need to investigate the reason behind this discrepancy. The relay uses the modified Takagi method given by (3.20) to calculate the distance to a fault. To understand the incorrect estimate from the relay during the line-to-line fault, let us start with the first source of fault-locating error, inaccurate fault current phasors. The accuracy of current phasors are affected when current transformers (CT) saturate or when fault currents have a decaying DC offset. The fault current phasors in this example are accurate since there is no evidence of CT saturation or DC offset in the waveforms shown in Figure 7.4. The second source of fault-locating error comes from inaccurate fault voltage phasors. We can rule out this error source as well since voltage waveforms show no evidence of transients. Furthermore, potential transformers are connected in wye and are measuring line-to-ground voltages. The next source of fault-locating error stems from inaccurate positive- and zero-sequence line impedance settings and is investigated next.

The phase and neutral conductors of this distribution feeder are known to be constructed using 336 ACSR conductor and 500 aluminum conductor, respectively. The characteristics of the conductor material are listed in Table 7.1. Line impedance calculation also needs knowledge about the spacing between the phase and neutral conductors. Since this information was not available, let us use the typical configuration of a 24.9 kV distribution feeder shown in Figure 7.6. Using Carson's equations and a 100 Ω-m earth resistivity value, the positive- and zero-sequence line impedances were calculated to be $z_{1,new} = 0.31 + j0.63 \; \Omega/\text{mi}$ and $z_{0,new} = 0.55 + j1.73 \; \Omega/\text{mi}$, respectively. Notice that these line impedance parameters are different than those used by the relay for fault location in Figure 7.7. The new line parameters, $z_{1,new}$ and $z_{0,new}$, were used to recalculate the distances to the fault. As seen

Table 7.1 Conductor Data [45].

	Material	Resistance (Ω/mi)	GMR (ft)
Phase	336 ACSR	0.306	0.0244
Neutral	500 AAC	0.206	0.0260

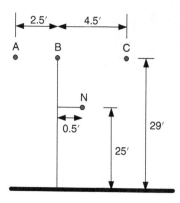

Figure 7.6 Typical configuration of a 24.9 kV distribution feeder [45].

```
Current Transformer Ratio (CTR) = 240
Potential Transformer Ratio (PTR) = 60
Positive-sequence Line Impedance (Z₁) = 0.28 + j0.54 Ω/mi
Zero-sequence Line Impedance (Z₀) = 0.61 + j2.07 Ω/mi
```

Figure 7.7 Relay settings.

Table 7.2 Fault Location with Revised Line Parameters.

Event	Fault Type	Actual Location (mi)	Estimated Location (mi)
1	BC		4.72
2	B-G		4.58
3	BC	4.43	4.54
4	B-G		4.58
5	BC-G		4.57

in Table 7.2, location estimates for all events are now close to the reported fault location. These findings strongly suggest that the disparity in location estimates was due to inaccurate line impedance settings. The utility should recalculate the line impedance and revise those settings in the relay.

The analysis above leads to a few more questions. For example, if the line impedance parameters were not correct, how did the modified Takagi method accurately locate the B-G fault but get compromised when locating BC or BC-G faults? After all, line impedance parameters are required to locate single line-to-ground faults as well. To answer the above questions, substitute V_G, I_G, and I_{seq} for a BC or a BC-G fault in (3.20) as shown below. Writing the resulting equation as a function of only the line impedance, we get the following expression:

$$m = \frac{\text{imag}\left(V_G \times I_{seq}^*\right)}{\text{imag}\left(z_1 \times I_G \times I_{seq}^*\right)}$$

$$= \frac{\text{imag}\left(\left(V_{G_b}^f - V_{G_c}^f\right) \times j I_{G_2}^{f*}\right)}{\text{imag}\left(z_1 \times \left(I_{G_b}^f - I_{G_c}^f\right) \times j I_{G_2}^{f*}\right)}$$

$$= \frac{\text{Constant}}{\text{imag}\left(z_1 \times \text{Constant}\right)} \quad [\text{mi}].$$

You can see that the distance to the fault calculation for a BC or a BC-G fault is inversely proportional to the positive-sequence line impedance, z_1. In this example, z_1 in the relay was set at a lower value than what is typical for a 24.9 kV distribution feeder. As a result, the Takagi method overestimated the distance to the fault.

To analyze the impact of line impedances on the fault location calculations for a single line-to-ground fault, substitute V_G, I_G, and I_{seq} for a B-G fault in (3.20). Writing the resulting expression as a function of the line impedance, we get

$$m = \frac{\text{imag}\left(V_G \times I_{seq}^*\right)}{\text{imag}\left(z_1 \times I_G \times I_{seq}^*\right)}$$

$$= \frac{\text{imag}\left(V_{G_b}^f \times I_{G_2}^{f*}\right)}{\text{imag}\left(z_1 \times \left(I_{G_b}^f + k I_{G_0}^f\right) \times I_{G_2}^{f*}\right)}$$

$$= \frac{\text{imag}\left(\text{Constant}\right)}{\text{imag}\left(\left(z_1 \times \text{Constant} + z_0 \times \text{Constant}\right) \times \text{Constant}\right)} \quad [\text{mi}].$$

Because the fault involves a path through ground, the distance to the fault calculation for a B-G fault is inversely proportional to both z_1 and z_0. Line impedance z_1 in the relay is set lower than the z_1 of a typical 24.9 kV feeder while z_0 is set greater than the z_0 of a typical 24.9 kV feeder. Coincidentally, the errors due to inaccurate z_1 and z_0 tend to cancel each other out. As a result, in spite of using incorrect line impedances, the relay in this example got lucky and located the B-G fault without a significant loss in accuracy.

7.3 Detect Instrument Transformer Installation Errors

Commissioning of protective relays and monitoring devices is an important step before placing them in service. The purpose of commissioning is to verify that the device has been installed and wired correctly, that it has been programmed correctly, and that it is operating as expected. During commissioning, one of the many checks that commissioning engineers perform is the verification of primary and secondary AC wiring [112]. This check verifies the installation of instrument transformers and ensures that the secondary AC wiring from the instrument transformers to protective relays or monitoring devices match the drawings. Examples of some common CT issues that get caught during commissioning are incorrect CT tap, incorrect CT polarity, and rolled CT wires as shown in Figure 7.8.

Once the installation and physical connections are verified, primary injection tests are carried out to further prove that there are no wiring errors and that the system works as expected. In a primary injection test, primary level currents and voltages are injected on the primary side of the electrical system. Performing this test requires using a small portable generator or a station service transformer. Several measurements are taken on the relay or

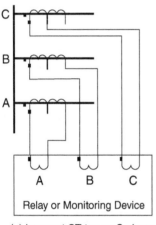

(a) Incorrect CT tap on C-phase.

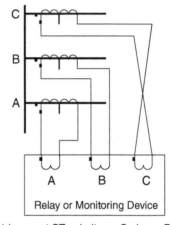

(b) Incorrect CT polarity on C-phase. Polarity got swapped at the device terminal.

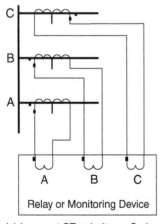

(c) Incorrect CT polarity on C-phase. CT installed with incorrect polarity.

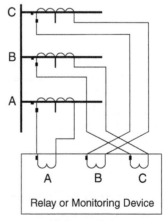

(d) Crossed phase.

Figure 7.8 Common CT installation and wiring errors.

the monitoring device. If the measured currents and voltages match the injected values, then the installation is correct and ready to be energized.

Despite these checks in place, some errors may go unnoticed during commissioning. The reason for such misses could be human error, shortcuts taken at the time of testing, or an incomplete commissioning checklist. The errors will find their way into service and may remain unnoticed for decades until a misoperation occurs and brings the error to light. Referring back to Figure 7.2, you can see that 9 percent of misoperations were due to as-left personnel error. Reference [113] presents examples of some frequently observed testing and commissioning errors and provides simple guidelines on how they can be avoided.

Incorrect fault location can help reveal incorrect setup of current and potential transformers and wiring errors that were missed during commissioning. Results of the analysis can be used to take corrective action and avoid future misoperations. This is demonstrated with the following example. Consider the 161 kV system shown in Figure 7.9. Station 1 and station 2 are connected by a 23.39-mile-long transmission line that has a positive- and zero-sequence impedance of $Z_1 = 2.85 + j18.22 \, \Omega$ primary and $Z_0 = 16.80 + j60.89 \, \Omega$ primary, respectively. The line experienced a three-phase fault due to a lightning strike at 5.86 miles from station 1 (17.53 miles from station 2). A digital fault recorder (DFR) at station 1 captured the voltage and current waveforms during the fault at 100 samples per cycle and is shown in Figure 7.10. Notice that the phase A current waveform at station 1 is missing. The spike in the A-phase voltage at the time of fault initiation backs up the theory that lightning was the root cause of the fault. Normally faults will cause the corresponding phase voltage to sag, not swell. Here, the current injected by lightning is responsible for the overvoltage condition. The prefault current at station 1 is 150 A while the fault current is 11 kA. An DFR at station 2 also captured the voltage and current waveforms during the fault at a sampling rate of 96 samples per cycle and is shown in Figure 7.11. The prefault current is 200 A while the fault current magnitude is 3.6 kA.

When we apply the simple reactance method to station 1 data, the distance-to-fault estimate is close to the actual location of the fault as seen in Table 7.3. Estimate from the same simple reactance method when applied to station 2 data is also close to the actual fault location. However, it is puzzling to observe that the distance estimate is negative. The most likely culprit is CT polarity. Either the CT had been installed with a reverse polarity as shown in Figure 7.12 or the polarity got swapped at the device terminal. The reverse CT polarity is further evident if one compares the fault currents at station 1 with that of station 2. Look at C-phase as an example. During the first half cycle of the fault, IC had a

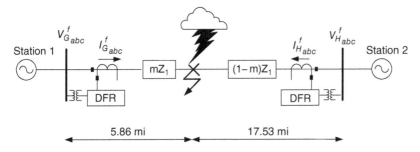

Figure 7.9 ABC fault at 5.86 miles from station 1 (17.53 miles from station 2).

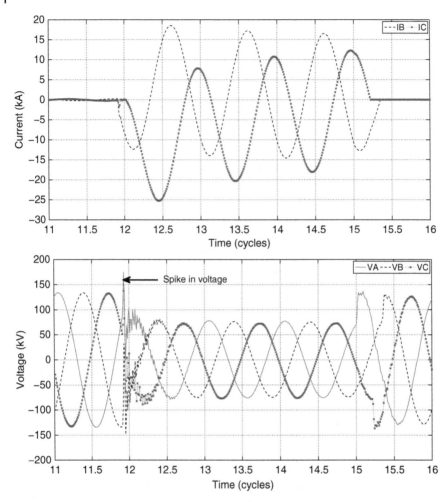

Figure 7.10 Waveform data captured by a DFR at station 1.

peak in the negative y-axis at station 1. However, at station 2, IC had a peak in the positive y-axis during the first half cycle of the fault. In other words, the current at station 1 was 180 degrees out of phase with the current at station 2 during the fault. This relationship is not expected during an internal fault. In fact, during an internal fault, the currents at both line ends should almost be in phase with each other. This further confirms that the CT polarity at station 2 had indeed been reversed and that the negative location estimate can be interpreted as 17.80 miles upstream with respect to the station 2 CT direction, as shown in Figure 7.12.

In this example, we demonstrate that incorrect fault location results should not be ignored or discarded. Getting to the root cause of the error can reveal an incorrect setup of power system equipment or wiring errors that got missed during field commissioning tests. This will alert utility personnel to take immediate action so as to avoid costly misoperations in the future. The CT at station 2 was installed with an incorrect polarity. As a result, a forward

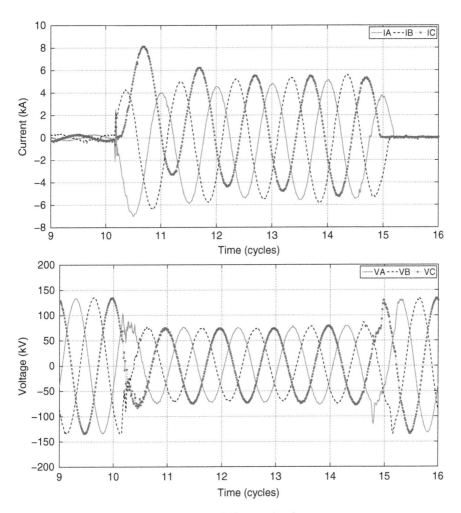

Figure 7.11 Waveform data captured by a DFR at station 2.

Table 7.3 Location Estimate from the Simple Reactance Method.

Station	Actual Location (mi)	Estimated Location (mi)
1	5.86	5.96
2	17.53	−17.70

fault appeared to be in the reverse direction. This can affect the reliability and performance of directional relays. Furthermore, the phase A current at station 1 was missing. It is possible that the phase CT has not been connected to the DFR and can result in loss of valuable information.

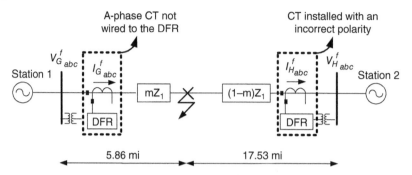

Figure 7.12 Negative fault location estimate using data from station 2 indicates reverse CT polarity.

7.4 Validate Zero-Sequence Line Impedance

Zero-sequence line impedance is a user-defined setting in relays and plays an important role in the correct operation of distance and directional relays. It is also required by impedance-based fault location algorithms to locate single line-to-ground faults. The zero-sequence line impedance is typically calculated using Carson's equations. Solving the equation, however, requires knowledge about earth resistivity, which is difficult to measure and changes with soil type, temperature, and moisture content in soils, as discussed in Chapter 4. A typical earth resistivity value used is $100\,\Omega$-m. If the actual earth resistivity deviates from this value, the actual zero-sequence line impedance will be quite different to the setting in the relay. Consequently, this will have a detrimental effect on system protection and fault location and efforts must be taken to validate the zero-sequence line impedance setting.

Chapter 2 discusses different techniques for determining the line impedance. One option is to inject signals and measure the line impedance. This requires scheduling an outage and having a generator source readily available. Another option is to calculate the zero-sequence line impedance using voltage and current signals at either end of the line when one pole of the breaker at one end of the line is open. Note that this intentional unbalance is necessary to measure the zero-sequence line impedance. This section presents two approaches that use the known distance to fault to circle back and validate the zero-sequence line impedance setting of the relay. The first approach, also known as the one-ended method, uses data from one end of a two-terminal transmission line or from a point upstream from the fault in a distribution feeder. The second approach, also known as the two-ended method, is applicable to two-terminal transmission lines only and uses data from both ends of the line. Both approaches are valid only when the fault has a return through the ground, i.e., during a single or a double line-to-ground fault.

One-Ended Method

This method validates the zero-sequence impedance of two-terminal transmission lines by using voltage and current phasors recorded at one end of the line during a ground fault. The ground fault can be either a single line-to-ground fault or a double line-to-ground fault. In addition to two-terminal lines, this method can also validate the zero-sequence impedance of radial transmission lines by using fault data captured at a terminal upstream from the ground fault. One-ended methods are easy to implement as they require data from only one terminal. However, the method assumes that the fault has zero resistance, that the fault type and fault location are known accurately, that the line is homogeneous, and that there is no mutual coupling. The method can also be applied to validate the zero-sequence impedance of distribution feeders but with some caution as distribution feeders are more non-homogeneous than transmission lines. The non-homogeneity can affect the accuracy of the line impedance estimate. The expressions for estimating the zero-sequence line impedance during a single line-to-ground and double line-to-ground fault are given below.

(a) Single Line-to-Ground Fault

Consider the two-terminal network sketched in Figure 3.1. Suppose that an A-phase-to-ground fault occurs at m per unit distance from terminal G. We start by writing (3.4) below, which derives an expression for calculating the voltage phasor of the faulted phase at the fault point, $V_{F_a}^f$, using terminal G data.

$$V_{F_a}^f = V_{G_a}^f - mZ_1 \left(I_{G_1}^f + I_{G_2}^f \right) - mZ_0 I_{G_0}^f \quad \text{[V]}.$$

If the fault is assumed to have a zero fault resistance, $V_{F_a}^f$ can be set equal to zero. The zero-sequence line impedance can then be estimated as

$$Z_0 = \frac{V_{G_a}^f - mZ_1 \left(I_{G_1}^f + I_{G_2}^f \right)}{m I_{G_0}^f} \quad \text{[Ω]}. \tag{7.1}$$

Generalizing the above equation for a single line-to-ground fault on any phase, we get:

$$Z_0 = \frac{V_G - mZ_1 \left(I_{G_1}^f + I_{G_2}^f \right)}{m I_{G_0}^f} \quad \text{[Ω]}, \tag{7.2}$$

where V_G is the voltage of the phase experiencing the single line-to-ground fault and is defined in Table 3.1. Note that (7.2) assumes that the fault location and the positive-sequence line impedance are known with a high degree of accuracy.

(b) Double Line-to-Ground Fault

Suppose that a double line-to-ground fault occurs at a distance m per unit from terminal G in Figure 3.1. During this unbalanced fault, the positive-, negative-, and zero-sequence networks are connected in parallel at the fault point. We are more interested in the negative- and zero-sequence networks when deriving the expression for calculating the zero-sequence line impedance. As a result, we lump the positive-sequence network in Figure 7.13 and expand the negative- and zero-sequence networks. Note that we have assumed a bolted fault with zero fault resistance. The negative-sequence voltage at the fault point, $V_{F_2}^f$, is equal to the zero-sequence voltage at the fault point, $V_{F_0}^f$, as shown below:

$$V_{G_2}^f - mZ_1 I_{G_2}^f = V_{G_0}^f - mZ_0 I_{G_0}^f. \tag{7.3}$$

The zero-sequence line impedance can then be estimated as follows:

$$Z_0 = \frac{V_{G_0}^f - V_{G_2}^f + mZ_1 I_{G_2}^f}{m \times I_{G_0}^f} \quad [\Omega]. \tag{7.4}$$

It is important that we know the fault location and the positive-sequence line impedance with a high degree of accuracy.

Two-Ended Method

This method applies to two-terminal transmission lines only. It calculates the zero-sequence line impedance using voltage and current phasors recorded at both terminals of a transmission line during a ground fault. This method is superior to the one-ended method as it makes no assumptions about the fault resistance. In other words, the accuracy of this method does not deteriorate when the fault has some resistance. There is no need to know whether the fault is a single line-to-ground fault or a double line-to-ground fault. The only requirement is that the fault involve a path through ground. Furthermore, this method does not require the user to know the fault location. If the location is unknown, it can be estimated using the two-ended methods described in Chapter 3. The method does assume that the line is homogeneous and that there is no mutual coupling. Depending on whether the recording devices at both ends of the line are GPS-time synchronized or not, the approaches below have been developed to calculate the zero-sequence line impedance.

(a) Two-Ended Synchronized Method

Use this method when the recording devices at both ends of the line are GPS-time synchronized. Consider only the zero-sequence network from Figure 7.13. Calculate $V_{F_0}^f$ from either

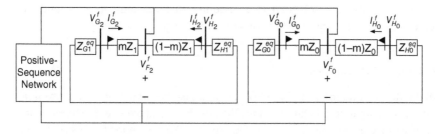

Figure 7.13 Connection of sequence networks during a double line-to-ground fault with zero fault resistance. We lump the positive-sequence network and expand the negative- and zero-sequence networks.

line end as shown below.

$$\text{Terminal G: } V_{F_0}^f = V_{G_0}^f - mZ_0 I_{G_0}^f,$$
(7.5)

$$\text{Terminal H: } V_{F_0}^f = V_{H_0}^f - (1 - m)Z_0 I_{H_0}^f.$$
(7.6)

Equating the above equations, we can solve for the zero-sequence line impedance as

$$Z_0 = \frac{V_{G_0}^f - V_{H_0}^f}{mI_{G_0}^f - (1 - m)I_{H_0}^f}.$$
(7.7)

(b) Two-Ended Unsynchronized Method

This method validates the zero-sequence impedance of two-terminal transmission lines when the recording devices at either end of the line are not GPS-time synchronized or have different sampling rates. The first step is to adjust the voltage and current phasors at terminal G with respect to those at terminal H. We use the negative-sequence network to calculate a synchronizing operator, $e^{j\delta}$, using the expression derived in (3.31). The equation is listed below for reference.

$$e^{j\delta} = \frac{V_{H_2}^f - (1 - m)Z_1 I_{H_2}^f}{V_{G_2}^f - mZ_1 I_{G_2}^f}$$

The new set of zero-sequence voltage and current phasors at terminal G that are synchronized with those at terminal H are $V_{G_0}^f e^{j\delta}$ and $I_{G_0}^f e^{j\delta}$. Using the same principle as that used in the two-ended synchronized method described above, the zero-sequence line impedance can be calculated as:

$$Z_0 = \frac{V_{G_0}^f e^{j\delta} - V_{H_0}^f}{mI_{G_0}^f e^{j\delta} - (1 - m)I_{H_0}^f}.$$
(7.8)

7.5   Calculate Fault Resistance

Voltage and current waveforms captured by IEDs during a fault can be used to estimate the magnitude of fault resistance and gain insight into the root cause of a fault. Reference [115] analyzed 148 fault events and found that most faults have fault resistance less than twenty ohms. Those with impedance greater than twenty ohms were caused by trees with a small diameter. It sounds counterintuitive but trees with a smaller diameter present more fault resistance than trees with a larger diameter [71]. Fault resistance alone, however, could not distinguish between animal contact- and lightning-induced faults. In addition to identifying the root cause of the fault, fault resistance also plays an important role in replicating the exact fault scenario in the circuit model and confirming the accuracy of the circuit model and in evaluating the performance of a distance relay. The Eriksson and Novosel et al. algorithms described in the previous chapters can be used to estimate the fault resistance from the waveform data captured at one end of the line as

$$R_f = \frac{d - mb}{f}.$$
(7.9)

The form taken by constants b, d, and f depends on the fault type and the number of terminals in a transmission line. For example, if a fault occurs on a one-terminal transmission line, constants b, d, and f are the same as those defined for the Novosel et al. algorithm in Section 5.1.7. For a fault on a two-terminal transmission line, constants b, d, and f are the same as those defined for the Eriksson algorithm in Section 3.1.4. It should be noted that the fault resistance is assumed to remain constant during the cycle used for calculating the voltage and current phasors.

7.6 Prove Short-Circuit Model

Utility engineers usually have detailed models of their transmission and distribution networks in a commercial power system software. The circuit model is useful for conducting short-circuit studies, determining protective relay settings, and choosing the maximum rating of circuit breakers and other power system equipment. Incorrect short-circuit model parameters can lead to erroneous relay settings and relay misoperations, an example of which is described in [116]. As a result, it is essential that the system model be accurate and be updated continually to reflect any system additions, repair, or modifications.

It is possible to validate the system short-circuit model if the fault location and fault resistance values are known. The same fault can be replicated at the known fault location in the short-circuit model. A good match between the current predicted by the short-circuit model and the actual fault current measurement confirms the accuracy of the circuit model. An example of this is shown in Figure 7.14. A B-G fault has occurred on a transmission line. The same fault was simulated in the system short-circuit model at the same fault location and with a fault resistance value that was estimated using was estimated using (7.9). The output from the short-circuit model was then compared against the fault

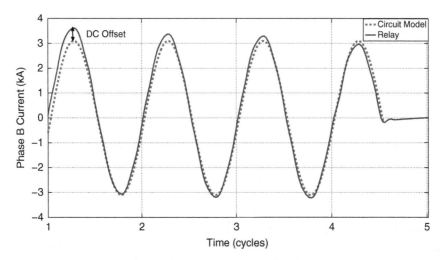

Figure 7.14 Short-circuit current output from the circuit model matched well with the fault current measured by a protective relay at the substation.

current measured by a protective relay at the substation during the system fault. Notice that there was some initial mismatch between the two currents. This is because the system fault current measured by the relay had some DC offset while the simulated current had none. However, as the DC offset started to decay, the two currents lined up quite well by the third cycle. This indicated that the short-circuit model was a good representation of the utility system.

7.7 Adapt Autoreclosing in Hybrid Lines

Many utilities have transmission lines that consist of both overhead and underground cable sections. They are known as hybrid lines (see Fig. 7.15). Hybrid lines introduce many challenges to line protection, one of them being autoreclosing. Autoreclosing is typically performed by a protective relay. When these devices detect a fault, they trip the circuit breaker and wait for a predefined time interval before issuing a close, giving the fault enough time to clear on their own. If the fault has cleared, all loads are automatically restored, which helps improve reliability. If the fault is still present, the relay may make several more attempts to clear the fault, depending on user settings. If all reclose attempts are unsuccessful, the fault is considered to be permanent, and the breaker is opened and locked out.

The majority of faults on overhead lines are temporary in nature. On such lines, autoreclosing can help improve reliability. However, faults on underground cable sections are more than likely permanent in nature. Autoreclosing on such faults would simply cause more damage and increase repair costs. So the question then becomes, should one autoreclose when a line has both overhead and underground sections, such as a hybrid line. One philosophy is to err on the conservative side and disable autoreclosing on hybrid lines. This has the obvious benefit of protecting the more expensive underground cable but hurts power quality when a temporary fault occurs on the overhead section. Another philosophy is to reclose for all faults since faults on underground cable sections are less frequent than overhead sections.

Knowledge about the fault location can provide a simple yet elegant solution to this problem. Recall that the traveling-wave fault location algorithm discussed in Chapter 3 is extremely accurate and has the ability to pinpoint the exact location of the fault. If the user inputs the exact length of the overhead and underground sections, the fault locator using the traveling-wave fault location algorithm can determine whether the fault is on the overhead or in the underground section. This result can be an input to a relay that is performing the autoreclose function. If the location is determined to be on the overhead

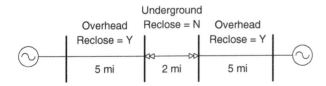

Figure 7.15 An example hybrid line.

Figure 7.16 Relay with the ability to send an adaptive autoreclose signal to a reclosing relay. The relay also has the ability to monitor the line for weak spots. (*Photo: Courtesy of Schweitzer Engineering Laboratories*).

section, autoreclosing would proceed as usual in an attempt to clear the fault. If the location is determined to be on the underground section, autoreclosing would be blocked to save the underground cable [63]. In other words, knowledge about the fault location makes the decision to autoreclose adaptive and not absolute. Figure 7.16 shows a relay that has adaptive autoreclose logic for hybrid lines.

Historically, accuracy has always been the fundamental requirement of fault location algorithms. Computational speed did not matter as long as the fault location output was accurate and available to maintenance personnel before dispatch. Distance to fault was usually calculated after the fault had already occurred and isolated by opening of the breaker. However, for innovative applications such as the one described in this section, the requirements change. Now, fault location needs to be calculated at protection speed, even before the breaker has opened, and needs to have the highest accuracy so that the device initiating the autoreclose can make the correct reclose decision.

7.8 Detect the Occurrence of Multiple Faults

When a recloser goes through its trip-reclose cycle and eventually locks out, it is not uncommon to assume that one permanent fault was responsible for all the trips. But that may not be the case all the time. Reference [54] describes a scenario in which a tree limb was responsible for creating the first fault. On reclose, it is possible that transient over-voltage caused a weak insulator string to flash over, creating a second fault on a different phase and at a different location. Fault location can help the utility detect such additional faults.

Another phenomenon that can create additional faults is conductor slap. A conductor slap occurs when the initial fault is a line-to-line fault. During this fault, the currents flowing through the faulted phases are equal in magnitude but out of phase with each other. The resultant magnetic forces are such that the two faulted conductors, all the way from the substation to the fault, move as far away from each other as possible as shown in Figure 7.17. When the fault is interrupted, the two conductors may contact one another during the return swing, causing a second fault upstream from the first fault.

Figure 7.17 Illustration of conductor slap.

(a) Conductors with sag.

(b) Faulted phase conductors move away from each other during a downstream line-to-line fault.

(c) After the fault is interrupted, the two phase conductors may make contact during the return swing, creating another fault upstream from the original fault.

The magnetic force between the two phase conductors during a line-to-line fault is given below [117]:

$$F = M \times \left(\frac{5.4 \times I^2}{S \times 10^7} \right),$$

(7.10)

where

F = pounds per foot of the conductor.

M = factor that accounts for DC offset. M = 8 for fully offset and 2 for no offset.

I = symmetrical fault current magnitude.

S = spacing between conductors in inches.

Maximum movement occurs during the following conditions:

1) Fault current magnitude. Higher fault current results in higher forces.
2) DC offset. This depends on the system X/R ratio and at what point of the voltage wave-form the fault occurred. A fully offset current produces four times more force than a non-offset current.
3) Spacing between conductors. Closer spacing produces higher forces.
4) Span length. The force equation above is in terms of pounds per foot of the conductor. Longer span lengths will produce larger forces.
5) Conductor sag. Spans with greater sag permit more movement.
6) Fault clearing time. If the fault clearing time is longer, i.e., the fault current flows through the conductors for a longer time, the conductors will have time to move farther away from each other as opposed to a shorter clearing time.

Fault location can help utilities identify the span that is experiencing the conductor slap event and take action to prevent such faults from reappearing in the future and stressing the system. Action items could include installing spacers that effectively cut the span length to half. Another option is to reevaluate the protection system and reduce fault clearing times. An online calculator at http://distributionhandbook.com/calculators/ asks the user to enter the expected fault current level, fault clearing time, conductor type, span length, sag, and the separation between conductors, as shown in Figure 7.18. The calculator will let the user know when the conductors will slap and what critical fault clearing time may prevent this slap. Other options include increasing the spacing between phase conductors, decreasing the span length by installing additional poles, and getting rid of slack. These last couple of options are more expensive to correct in an existing installation but should be considered during new construction.

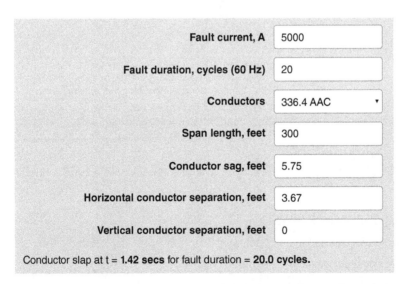

Fault current, A	5000
Fault duration, cycles (60 Hz)	20
Conductors	336.4 AAC ▾
Span length, feet	300
Conductor sag, feet	5.75
Horizontal conductor separation, feet	3.67
Vertical conductor separation, feet	0

Conductor slap at t = **1.42 secs** for fault duration = **20.0 cycles.**

Figure 7.18 Online calculator that shows when conductor slap will occur and what critical fault clearing time may prevent it.

(a) Initial line-to-line fault downstream from relay 2 due to fallen tree branch. Faulted phase conductors between the substation to the fault point move away from each other.

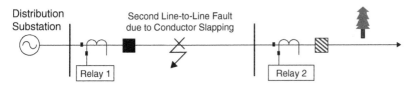

(b) When relay 2 opens the breaker and interrupts the fault, faulted phase conductors upstream from the original fault may make contact during the return swing and create a second fault.

(c) Relay 1 opens the breaker to clear the second fault. Line patrol find evidence for the original fault (tree branch) and suspect miscoordination between relay 1 and relay 2.

Figure 7.19 Operation of relay 1 due to conductor slap can be mistaken for miscoordination.

In addition to stressing the system, conductor slap can lead utilities on a wild goose chase as the operation of an upstream breaker due to conductor slap can often be mistaken for miscoordination issues [118]. For example, suppose that a tree branch causes a line-to-line fault downstream from relay 2 as shown in Figure 7.19. The two faulted conductors move as far away from each other as possible. When relay 2 detects and isolates the fault, the two conductors will attempt to come back to their original position. However, on this return swing, they may overshoot and contact each other or come close enough to break down air and cause a flashover. If this contact happens to occur in the protection zone of the upstream relay (relay 1 in this example), the upstream device will operate and clear the fault. The initial fault is typically always found. But the subsequent fault due to conductor slap is harder to find during line patrol. This may lead utilities to suspect miscoordination between relay 1 and relay 2. A lot of time and resources will be spent reviewing coordination settings when in fact there were two separate faults and both relays operated correctly. Fault location can help end this confusion.

7.9 Identify Impending Failures and Take Corrective Action

Temporary faults such as those caused by flashover due to dirty insulators, encroaching vegetation, failing power system equipment, or clashing of conductors due to strong winds can be difficult to find. Fault location can help linemen find those temporary faults and take corrective action before they develop into a permanent fault and cause an outage. Corrective actions could include washing the insulator, cutting down vegetation to improve clearance, repairing power system equipment, or installing phase spacers. Reference [119] describes an incident in which a failing transformer bushing gave a six-week notice about an impending failure. During those six weeks, it flashed over five times. Each time, the flashover was successfully cleared by an unmonitored recloser. None of the customers complained about the temporary blip in power. As a result, the utility had no idea about this looming failure until the bushing failed permanently and caused a power outage. It is not possible to avoid faults altogether. However, by tracking temporary faults and by taking note of the early warning signs, it is possible to prevent recurring faults, reduce the number of permanent faults, and improve reliability and quality of service. Figure 7.16 shows an example of a relay that can monitor the line for weak spots and flag recurring events.

7.10 Exercise Problems

Time to get to work! In this section, you will work through some exercise problems on validating the zero-sequence line impedance. You will need a calculator.

■ Exercise 7.1

Refer to the one-line given in Figure 7.20. An arrester failed and created a single line-to-ground fault on phase A of a 161 kV transmission line. The transmission line that experienced the fault is 21.15 miles long and connects terminal G with terminal H. The positive- and zero-sequence impedances of the line are $Z_1 = 3.18 + j16.68$ Ω primary and $Z_0 = 15.21 + j52.45$ Ω primary, respectively. The failed arrester is located 14.90 miles from terminal G (6.25 miles from terminal H). Digital fault recorders (DFRs) at both terminals recorded the three-phase line-to-ground voltages and three-phase currents during the fault at 100 samples per cycle. They are shown in Figure 7.21 and Figure 7.22. The DFRs are not synchronized to a GPS time source. A Fourier filter was used to extract the terminal G voltage and current phasors during the fault (cycle 8.5) to be:

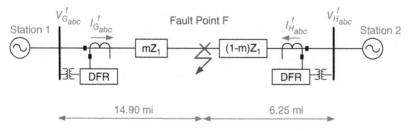

Figure 7.20 Exercise 7.1: A-G fault located 14.90 miles from terminal G.

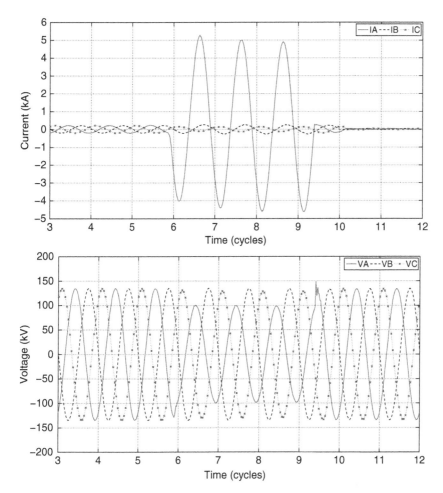

Figure 7.21 Exercise 7.1: DFR measurements at terminal G during the A-G fault caused by a failed arrester.

$$I^f_{G_{abc}} \text{ [A]} \qquad\qquad V^f_{G_{abc}} \text{ [kV]}$$

$$\begin{bmatrix} 3352.30\angle - 110.21° \\ 176.10\angle - 143.98° \\ 96.4\angle 83.11° \end{bmatrix} \qquad \begin{bmatrix} 69.81\angle - 36.71° \\ 94.90\angle - 152.34° \\ 91.16\angle 85.05° \end{bmatrix}.$$

The Fourier filter was also used to extract the terminal H voltage and current phasors during the fault (cycle 20.5) to be:

$$I^f_{H_{abc}} \text{ [A]} \qquad\qquad V^f_{H_{abc}} \text{ [kV]}$$

$$\begin{bmatrix} 6065.50\angle - 113.44° \\ 185.90\angle 33.98° \\ 73.80\angle - 96.98° \end{bmatrix} \qquad \begin{bmatrix} 53.88\angle - 38.57° \\ 98.10\angle - 158.74° \\ 96.13\angle 87.53° \end{bmatrix}.$$

Given the above data, validate the zero-sequence line impedance.

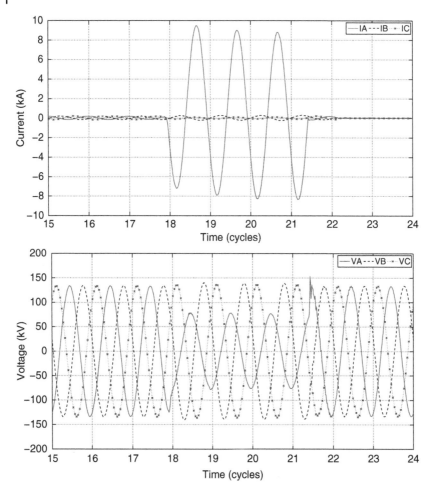

Figure 7.22 Exercise 7.1: DFR measurements at terminal H during the A-G fault caused by a failed arrester.

Solution:

Let us review the data we have. Actual fault location is known. Voltage and current phasors at both ends of the line during the ground fault are available. The phasors are not synchronized. Based on this, we will be using the two-ended unsynchronized method. We could have used the one-ended approach as well. However, we chose the two-ended method as it will give us better accuracy over the one-ended method since the two-ended method makes no assumptions about the fault being bolted in nature.

When implementing the two-ended method, we need to first calculate the negative-sequence voltage and current at terminal G and terminal H. Since this is an A-G fault, we use the A-phase as the reference. We don't have the phase rotation data. However, this can be easily figured out by looking at the prefault data. In Figure 7.21, look at the voltage phasors starting from the fourth cycle. VA peaks first in the negative y-axis followed by VB followed by VC. Thus, the system phase rotation is ABC.

The sequence currents at terminal G are:

$$
\begin{bmatrix} I^f_{G_0} \\ I^f_{G_1} \\ I^f_{G_2} \end{bmatrix} = \frac{1}{3} \begin{bmatrix} 1 & 1 & 1 \\ 1 & a & a^2 \\ 1 & a^2 & a \end{bmatrix} \times \begin{bmatrix} I^f_{G_a} \\ I^f_{G_b} \\ I^f_{G_c} \end{bmatrix}
$$

$$
= \frac{1}{3} \begin{bmatrix} 1 & 1 & 1 \\ 1 & a & a^2 \\ 1 & a^2 & a \end{bmatrix} \times \begin{bmatrix} 3352.30\angle -110.21° \\ 176.10\angle -143.98° \\ 96.4\angle 83.11° \end{bmatrix}
$$

$$
= \begin{bmatrix} 1135.60\angle -112.23° \\ 1134.00\angle -105.69° \\ 1087.90\angle -112.81° \end{bmatrix} \quad [A].
$$

The sequence voltages at terminal G are:

$$
\begin{bmatrix} V^f_{G_0} \\ V^f_{G_1} \\ V^f_{G_2} \end{bmatrix} = \frac{1}{3} \begin{bmatrix} 1 & 1 & 1 \\ 1 & a & a^2 \\ 1 & a^2 & a \end{bmatrix} \times \begin{bmatrix} V^f_{G_a} \\ V^f_{G_b} \\ V^f_{G_c} \end{bmatrix}
$$

$$
= \frac{1}{3} \begin{bmatrix} 1 & 1 & 1 \\ 1 & a & a^2 \\ 1 & a^2 & a \end{bmatrix} \times \begin{bmatrix} 69.81\angle -36.71° \\ 94.90\angle -152.34° \\ 91.16\angle 85.05° \end{bmatrix}
$$

$$
= \begin{bmatrix} 6.95\angle 165.99° \\ 85.25\angle -34.46° \\ 8.99\angle 147.50° \end{bmatrix} \quad [kV].
$$

The sequence currents at terminal H are:

$$
\begin{bmatrix} I^f_{H_0} \\ I^f_{H_1} \\ I^f_{H_2} \end{bmatrix} = \frac{1}{3} \begin{bmatrix} 1 & 1 & 1 \\ 1 & a & a^2 \\ 1 & a^2 & a \end{bmatrix} \times \begin{bmatrix} I^f_{H_a} \\ I^f_{H_b} \\ I^f_{H_c} \end{bmatrix}
$$

$$
= \frac{1}{3} \begin{bmatrix} 1 & 1 & 1 \\ 1 & a & a^2 \\ 1 & a^2 & a \end{bmatrix} \times \begin{bmatrix} 6065.50\angle -113.44° \\ 185.90\angle 33.98° \\ 73.80\angle -96.98° \end{bmatrix}
$$

$$
= \begin{bmatrix} 1993.60\angle -112.28° \\ 2015.10\angle -112.88° \\ 2059.50\angle -112.17° \end{bmatrix} \quad [A].
$$

The sequence voltages at terminal H are:

$$\begin{bmatrix} V^f_{H_0} \\ V^f_{H_1} \\ V^f_{H_2} \end{bmatrix} = \frac{1}{3} \begin{bmatrix} 1 & 1 & 1 \\ 1 & a & a^2 \\ 1 & a^2 & a \end{bmatrix} \times \begin{bmatrix} V^f_{H_a} \\ V^f_{H_b} \\ V^f_{H_c} \end{bmatrix}$$

$$= \frac{1}{3} \begin{bmatrix} 1 & 1 & 1 \\ 1 & a & a^2 \\ 1 & a^2 & a \end{bmatrix} \times \begin{bmatrix} 53.88\angle - 38.57° \\ 98.10\angle - 158.74° \\ 96.13\angle 87.53° \end{bmatrix}$$

$$= \begin{bmatrix} 17.52\angle 149.25° \\ 82.59\angle - 36.27° \\ 11.33\angle 146.14° \end{bmatrix} \quad [\text{kV}].$$

The distance to fault from terminal G in per unit of the line length is

$$m = \frac{14.90}{21.15} = 0.70 \quad [\text{pu}].$$

Because data between terminal G and terminal H are not GPS-time synchronized, we calculate the synchronizing operator, $e^{j\delta}$, using (3.31) as:

$$e^{j\delta} = \frac{V^f_{H_2} - (1-m)Z_2 I^f_{H_2}}{V^f_{G_2} - mZ_2 I^f_{G_2}}$$

$$= \frac{11.33\angle 146.14° \times 1000 - \left(0.3 \times (3.18 + j16.68) \times 2059.5\angle - 112.17°\right)}{8.99\angle 147.50° \times 1000 - 0.70 \times (3.18 + j16.68) \times 1087.9\angle - 112.81°}$$

$$= 1\angle - 0.28°.$$

Multiplying terminal G data with the synchronizing operator will synchronize terminal G with terminal H. Next we can use (7.8) to compute the zero-sequence line impedance as:

$$Z_0 = \frac{V^f_{G_0} e^{j\delta} - V^f_{H_0}}{m I^f_{G_0} e^{j\delta} - (1-m) I^f_{H_2}}$$

$$= \frac{\left((6.95\angle 165.99° \times 1\angle - 0.28°) - 17.52\angle 149.25°\right) \times 1000}{(0.70 \times 1135.60\angle - 112.23° \times 1\angle - 0.28°) - (0.30 \times 2059.50\angle - 112.17°)}$$

$$= 18.56 + j59.46 \quad [\Omega].$$

Recall that the setting in the relay was $Z_0 = 15.21 + j52.45 \ \Omega$.

■ **Exercise 7.2**

Consider the utility network shown in Figure 7.23. The rated voltage at substation A is 161 kV. A microprocessor-based digital relay protects the 23.6-mile-long radial transmission line that connects substation A with substation C. The arrangement of the phase and shield conductors is shown in Figure 7.24. The shield conductors protect the phase conductors from direct lightning strikes. Data for the phase and shield conductors are listed in Table 7.4. Using Carson's equations, the positive- and zero-sequence line impedances

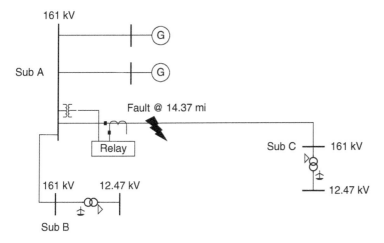

Figure 7.23 Exercise 7.2: Utility network.

Figure 7.24 Exercise 7.2: Overhead transmission line spacing in feet.

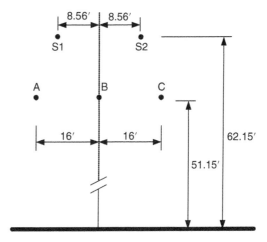

were calculated to be $Z_1 = 6.01 + j19.00\,\Omega$ and $Z_0 = 19.72 + j56.23\,\Omega$ primary, respectively. When a B-G occurred 14.37 miles from substation A, the relay at that substation detected and cleared the fault by opening up the substation breaker. It also recorded an event report that showed the voltage and current waveforms before and during the fault. This is shown in Figure 7.25. Notice that this was a short-duration fault that lasted for about four cycles before being cleared by the relay. It also had some DC offset toward the beginning of the fault. To minimize any error due to this offset, the third cycle after fault inception was chosen to calculate the current and voltage phasors during the fault. The phasors, extracted using the Fourier filter, are given below.

$$
I^f_{G_{abc}}\,[\text{A}] \qquad\qquad V^f_{G_{abc}}\,[\text{kV}]
$$

$$
\begin{bmatrix}
84.18\angle - 83.70° \\
2285.00\angle 90.11° \\
162.69\angle 65.77°
\end{bmatrix}
\qquad
\begin{bmatrix}
99.62\angle - 71.28° \\
46.74\angle 158.10° \\
102.66\angle 37.10°
\end{bmatrix}.
$$

Validate the zero-sequence line impedance.

Table 7.4 Exercise 7.2: Conductor Data.

	Material	Resistance (Ω/mi)	Diameter (inch)	GMR (feet)
Phase conductor	397,500 26/7 ACSR	0.2537	0.7836	0.0265
Shield wire	3/8 A HSS	5.6500	0.3600	0.0120

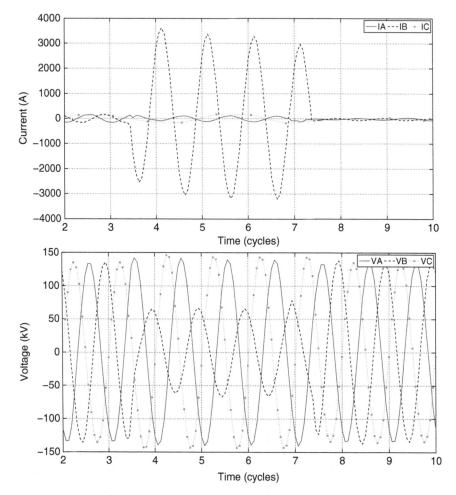

Figure 7.25 Exercise 7.2: Current and voltage waveforms during a B-G fault on a 161 kV transmission line. Root cause of the fault is unknown.

Solution:

It is possible to validate the zero-sequence line impedance as the fault described in this example involves a return path through the ground. Because the line is radial, the one-ended approach described in Section 7.4 was used. First, we need to calculate the sequence currents during the fault. We use B-phase as the reference. Looking at the voltage waveform during the second cycle (prefault), A-phase peaks in the negative y-axis followed by B-phase and then C-phase. This means that the system phase rotation is ABC.

$$\begin{bmatrix} I_{G_0}^f \\ I_{G_1}^f \\ I_{G_2}^f \end{bmatrix} = \frac{1}{3} \begin{bmatrix} 1 & 1 & 1 \\ 1 & a & a^2 \\ 1 & a^2 & a \end{bmatrix} \times \begin{bmatrix} I_{G_b}^f \\ I_{G_c}^f \\ I_{G_a}^f \end{bmatrix}$$

$$= \frac{1}{3} \begin{bmatrix} 1 & 1 & 1 \\ 1 & a & a^2 \\ 1 & a^2 & a \end{bmatrix} \times \begin{bmatrix} 2285.00\angle90.11° \\ 162.69\angle65.77° \\ 84.18\angle-83.70° \end{bmatrix}$$

$$= \begin{bmatrix} 783.59\angle88.26° \\ 771.77\angle96.04° \\ 736.18\angle85.89° \end{bmatrix} \quad [A].$$

The per unit distance to the fault can be calculated as:

$$m = \frac{14.37}{23.60} = 0.6089 \quad [\text{pu}].$$

Next we use (7.2) to calculate the zero-sequence line impedance as:

$$Z_0 = \frac{V_G - mZ_1 \left(I_{G_1}^f + I_{G_2}^f \right)}{m I_{G_0}^f}$$

$$= \frac{46.74\angle158.1° \times 1000 - 0.6089 \times (6.01 + j19) \times 1502.04\angle91.09°}{0.6089 \times 783.59\angle88.26°}$$

$$= 24.05 + j55.01 \quad [\Omega].$$

The estimated zero-sequence line impedance matches well with that computed using Carson's equations with an earth resistivity value of $100\,\Omega$-m.

7.11 Summary

This chapter presents the additional benefits of fault location for improving power system performance and reliability. Potential applications include assessing the performance of relays, discovering erroneous relay settings and CT/PT wiring errors, validating the zero-sequence line impedance setting, estimating fault resistance and gaining insight into the root cause of the fault, confirming the accuracy of the short-circuit model, detecting whether multiple faults occurred in a trip-close-lockout sequence, identifying spans susceptible to conductor slaps and taking corrective action, and tracking down temporary faults and preventing them from turning into a permanent fault. Finally, two exercises show how to validate the zero-sequence line impedance setting of in-service relays by using data captured during a ground fault.

A

Fault Location Suite in MATLAB

The theory behind several impedance-based fault-locating algorithms was discussed in Chapters 3 and 5, respectively. This appendix presents a fault location suite that uses waveform data captured by fault recorders and intelligent electronic devices (IEDs) during a fault on a line and determines the location of the fault.

Nowadays, single-ended impedance-based fault location is available as a standard feature in protective relays. However, relays can report erroneous fault location estimates due to inaccurate user settings (such as line impedance or line length) or challenging fault scenarios (such as those with varying fault resistance value). Hence, the ability to calculate the fault location manually is greatly beneficial for electric utility engineers. Furthermore, this fault location suite is a useful tool for university students and engineers to gain hands-on experience on the fault location estimation process from event reports using the algorithms studied in this book.

The complex calculations involved in applying the impedance-based fault location methods can now be done using any high-level technical computing platform and need not be performed entirely manually. MATLAB is one such software that can aid us in this process. MATLAB software, developed by MathWorks, is a mathematical computation environment with a programming language that can be used for analysis and visualization applications.

The fault location suite provided with this book can be downloaded from www.wiley .com/go/das/faultlocation. In this fault location suite, scripts written in MATLAB are used to preprocess fault record data, visualize the recorded current and voltage waveforms, and compute the fault location estimate. This chapter demonstrates the application of this tool by computing the location of an actual fault that occurred on a utility transmission feeder using various algorithms discussed earlier in this book.

A.1 Understanding the Fault Location Script

The fault location suite consists of a collection of scripts that require MATLAB to run and execute them. These scripts have been developed and tested using MATLAB version

This chapter was contributed by Dr. Sundaravaradan N. Ananthan.

Fault Location on Transmission and Distribution Lines: Principles and Applications, First Edition.
Swagata Das, Surya Santoso, and Sundaravaradan N. Ananthan.
© 2022 John Wiley & Sons Ltd. Published 2022 by John Wiley & Sons Ltd.
Companion website: www.wiley.com/go/das/faultlocation

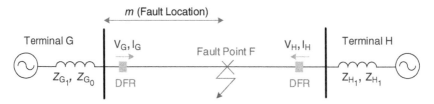

Figure A.1 Two-terminal line.

R2020a. The fault location suite can be used to locate single line-to-ground, line-to-line, double line-to-ground, and three-phase faults on a radial or two-terminal line. Figure A.1 shows a two-terminal line similar to Figure 3.1 with a fault at a distance m from terminal G. The final output is the fault location (m). This section walks the reader through the fault location process and explains the MATLAB scripts in the fault location suite with an example in the steps listed below.

- Step 1: Gather information required for the fault location process.
- Step 2: Gain an overview of the files in the fault location suite.
- Step 3: Enter the known data and system information.
- Step 4: Import waveform data recorded at the local and remote ends of the line.
- Step 5: Visualize waveform data and determine fault and prefault sample.
- Step 6: Apply fault location algorithms.
- Step 7: View the fault location estimates from the fault location algorithms.

The example scenario used to demonstrate the application of this tool is an actual fault that occurred on a utility transmission feeder. It is Exercise 7.1 in Chapter 7 of this book. Recapping, a 161 kV transmission line experienced a single line-to-ground fault on phase A caused by a failed line arrester. The transmission line is 21.15 miles long and connects terminals G with terminal H similar to Figure A.1. The positive- and zero-sequence impedances of the line are $Z_1 = 3.18 + j16.68\ \Omega$ primary and $Z_0 = 15.21 + j52.45\ \Omega$ primary, respectively. The failed line arrester was located at 14.90 miles from terminal G, or 6.25 miles from terminal H. The DFRs at both stations recorded the three-phase line-to-ground voltage and three-phase current waveforms during the fault at 100 samples per cycle. The DFRs were not synchronized to a GPS time source.

Step 1: Gather information required for the fault location process

- Fault Event Report
  The first and foremost item required for the fault location process is the fault event report. There are two types of event reports that are typically recorded by a relay or fault recorder. The first is a raw event report, which contains the exact instantaneous current and voltage values observed by the recording device. The recorded waveforms will include both the fundamental and harmonic components of the signal. The other type of event report is a filtered event report. The fundamental quantities are extracted by passing the measured signals through a filter and stored in this event report. To perform fault location, we extract steady-state fundamental fault current and voltage values and use them in the fault location algorithms. Hence, we use the filtered event report for this process.

Different digital fault recorders have different recording capabilities, specifications, and record in different file formats. The fault location suite is programmed to read data from a *.csv file. Fault records present in other standard file formats such as IEEE COMTRADE format [120] can be converted to *.csv file format using event report visualization and analysis software such as PQDiffractor, developed by Electrotek Concepts Inc. Another common event report file format is *.cev and can be converted to *.csv using Synchrowave Event software, developed by Schweitzer Engineering Laboratories.

The next required information to interpret the data in the event reports is identifying the units in which each of the channels were recorded. If the event report is from a relay, the relay's instruction manual can provide this information. For an event report in COM-TRADE file format, the configuration file contains this information required to interpret the contents of the data file. The sampling rate can also be found in the configuration file.

- Phase Rotation and Fault Type
  The phase rotation and fault type are critical inputs to impedance-based fault location algorithms, as discussed in Chapter 2. The phase sequence can be determined by observing the order in which the three-phase prefault voltages reach their respective peaks. For example, a system with ABC phase sequence will have the A-phase voltage reach its peak first, followed by the B-phase voltage, and then followed by the C-phase voltage.

  The fault type too can be identified by looking at the waveform data. For example, we can identify an event as an A-G fault if there was an increase in only the A-phase current and a decrease in only the A-phase voltage. The event report summary too presents the fault type declared by the relay. This should typically match with our fault type selection unless the fault evolved from one type to another in the same event report. In such cases, the fault type chosen should match with the section of the fault event waveform being used to calculate the fault location.

- Event Report from the Remote End of the Line (If Available)
  When event reports are available from both ends of a line, it is good to implement two-ended methods and take advantage of the various benefits they provide over one-ended methods. The fault location suite does not require event reports from both ends to be time aligned, be downloaded with the same resolution, or be captured by the same relay type, but they need to be filtered event reports. The event reports recorded by relays at each end of the line can be considered as synchronized if the relays have a common time source or the clocks between the two relays are synchronized and the trigger times for the relays to record the fault event are similar.

- Line Parameters
  The last set of information required is the line parameters such as line length and line impedance. The line length can be entered either in miles or kilometers. In this fault location suite, all the calculations are done in primary quantities. Hence, the line impedance must be entered in primary ohms. When the CT and PT ratios are known, the impedance can be converted from secondary ohms to primary ohms using the following equation.

$$Z_{py} = \frac{Z_{sec} Z_{base_{sec}}}{Z_{base_{py}}} = Z_{sec} \left(\frac{V_{base_{sec}}}{V_{base_{py}}} \right) \left(\frac{I_{base_{py}}}{I_{base_{sec}}} \right) \tag{A.1}$$

$$= Z_{sec} \frac{\text{CT Ratio}}{\text{PT Ratio}} \Omega. \tag{A.2}$$

The source impedance parameters are automatically calculated from the waveform data by the fault location suite.

Step 2: Gain an overview of the files in the fault location suite

The fault location suite folder contains the following files:

- Master Script
  The master script is the primary MATLAB script to calculate the fault location. The user interacts with the master script to input the available data, visualize the recorded waveforms, and view the estimated fault location.
- Fault Location Algorithms
  The master script calls the fault location algorithms function files to calculate the fault location.
- Supporting File
  The supporting file assists the master script to calculate the magnitude and phase angle of a phasor from the event report data using the fast Fourier transform.
- Fault Event Report Data of an Example Case
  There are two files in the fault location suite containing real-world fault data recorded by fault recorders at the two ends of a transmission line.

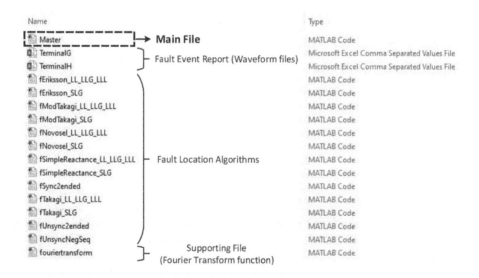

This fault location suite in MATLAB can implement the following fault location algorithms:

- One-Ended Impedance-Based Algorithms
  - simple reactance
  - Takagi
  - modified Takagi
  - Eriksson et al. (current distribution factor method)
  - Novosel et al.

- Two-Ended Impedance-Based Algorithms
  - synchronized
  - unsynchronized
  - unsynchronized negative-sequence

The algorithms listed above have been described in Sections 3.1, 3.2, and 5.1.

Step 3: Enter the known data and system information
The next steps from this point are carried out in the master script. The first section of the master script contains variables regarding the system information, as shown in the figure below. Enter the line length (in miles or kilometers), line impedance (in primary ohms), fault type, sampling frequency (in samples per cycle), and phase rotation of the system identified in Step 1.

```
Line_Length_units = 'miles';   % Line length units (miles or kilometers)
LL = 21.15;                    % Total line length
Z_1 = 3.1805 + 16.6750i;       % Positive-sequence line impedance in primary ohms
Z_0 = 15.2130 + 52.4537i;      % Zero-sequence line impedance in primary ohms

Fault_type = 'AG';             % Fault type (AG, BG, CG, AB, BC, CA, AB-G, BC-G, CA-G, ABC)

N_G = 100;                     % Sampling rate of the DFR at Terminal G in samples per cycle

Rotation = 'ABC';              % System Phase Rotation (ABC or ACB)
```

The script first takes in information about the event report recorded at terminal G (local end) and later takes the information for the event report recorded at terminal H (remote end) if available. In this event, the fault recorder identified the fault to be a single line-to-ground fault on phase A. The sampling rates of the recorders at both terminals (N_G and N_H) were 100 samples per cycle (or a sampling frequency of 6000 Hz). The final input information required in this set is the phase rotation and is ABC in our example.

Step 4: Import waveform data recorded at the local and remote ends of the line

To import the waveform data, change the working directory to the location in which the *.csv file is located and enter the file name. In this example, the filename of the event report recorded at terminal G is "TerminalG.csv" and the event report recorded at terminal H is "TerminalH.csv." Indicate the file name and the sheet name to be read in the master script as follows:

```
cd('C:\.....\Fault Location Suite');
filename = 'TerminalG.csv';
sheetname = 'TerminalG';
```

On running the master script, MATLAB prompts the user to enter the unit of the voltage and current waveforms, as shown below. The conversion factor is used to import the waveform data in kV and the current waveform data in kA. In this example, the voltage waveforms are already recorded in kV and the current waveforms are recorded in kA at both terminals G and H.

```
prompt = 'What is the unit of voltage waveforms recorded at Terminal G? (kV or V)\n ';
Unit_V_G = input(prompt,'s');
disp(' ');

prompt = 'What is the unit of current waveforms recorded at Terminal G? (kA or A)\n ';
Unit_I_G = input(prompt,'s');
disp(' ');

if strcmp(Unit_V_G,'kV')== 1
    Conversion_Factor_V_G = 1;
else
    Conversion_Factor_V_G = 0.001;
end

if strcmp(Unit_I_G,'kA')== 1
    Conversion_Factor_I_G = 1;
else
    Conversion_Factor_I_G = 0.001;
end
```

The time information is imported using the following code:

```
time_G = xlsread(filename,sheetname,'A2:A7019');
```

The voltage portion of the event report recorded at terminal G contains the instantaneous line-to-ground voltage measurements in all three phases. The following MATLAB code is used to import the voltage waveforms.

```
V_A_waveG = xlsread(filename,sheetname,'B2:B7019')*Conversion_Factor_V_G;
V_B_waveG = xlsread(filename,sheetname,'C2:C7019')*Conversion_Factor_V_G;
V_C_waveG = xlsread(filename,sheetname,'D2:D7019')*Conversion_Factor_V_G;
```

The current portion of the event report recorded at terminal G contains the instantaneous line current measurements in all three phases. The current waveforms are imported using the following code:

```
I_A_waveG = xlsread(filename,sheetname,'E2:E7019')*Conversion_Factor_I_G;
I_B_waveG = xlsread(filename,sheetname,'F2:F7019')*Conversion_Factor_I_G;
I_C_waveG = xlsread(filename,sheetname,'G2:G7019')*Conversion_Factor_I_G;
```

The above code to import waveforms should be edited appropriately to match the time, voltage, and current channels with their corresponding columns in the csv file. The row numbers too need to be updated with either the total number of samples recorded or the length of the waveform the user wishes to import into MATLAB. To import the entire waveform, use the last row number of the csv file.

The code then prompts the user asking whether the line being analyzed is a two-terminal line. If the user responds with a "Yes," it then asks the user if they have the event report recorded at the remote end of the line (terminal H) and whether the two event reports are synchronized.

```
prompt = 'Is this a two-terminal transmission line? Enter Yes or No. \n';
Two_terminal = input(prompt,'s');
disp(' ');

if strcmp(Two_terminal,'Yes')==1
    prompt = 'Do you have event report recorded at other end of the line (Terminal H)? (Yes
or No)\n ';
    Remote = input(prompt,'s');
    disp(' ');
end
```

```
if strcmp(Remote,'Yes')==1
    prompt = 'Are data from two terminals synchronized? (Yes or No) \n';
    Data_sync = input(prompt,'s');
    disp(' ');
end
```

If the event report from terminal H is available, update the MATLAB script to enter the file path, sampling frequency, and channel information. Run the script again and respond to the above query with a "Yes" to import the instantaneous line-to-ground voltages and line currents recorded at terminal H in units of kV and kA, respectively. After importing the data, the master script plots the three-phase voltage and current waveforms. Update the limits of the x and y axes of the plots accordingly to get a clear view of the fault waveforms.

Recollect from Section 3.1 and 5.1 that while the Eriksson et al. method is valid for locating faults on two-terminal transmission lines, the Novosel et al. method is applicable for locating faults on radial transmission lines only. The two-ended impedance-based algorithms cannot be applied to radial transmission lines. Furthermore, if measurements from both ends of a two-terminal transmission line are not synchronized, then the synchronized two-ended method cannot be applied. In this example, the faulted line is a two-terminal transmission line, and the measurement data from the DFRs at terminals G and H are not synchronized.

Step 5: Visualize waveform data and determine fault and prefault sample

Fault recorders capture instantaneous snapshots of the various quantities they measure and record them in the event reports. Hence, the data present is in the form of samples and need to be preprocessed to extract information in a usable format (phasors) for fault analysis.

Impedance-based fault location algorithms require the input of fundamental frequency voltage and current phasors before and during the fault to accurately calculate the location estimate. In this fault location suite, the fast Fourier transform (FFT) operation is employed to calculate the prefault and fault phasors. The FFT operation successfully discards all harmonics and extracts the magnitude and phase angle of only the fundamental frequency components. The length of the data window used here is one cycle at power frequency.

The FFT in this suite is implemented as a function in the fouriertransform.m script file, as shown below. The function extracts one cycle of waveform data starting from *samplestart* and performs FFT to extract the magnitude and phase angle of the fundamental frequency. Note that the recorded waveforms are the instantaneous values of voltages and currents. As the input to the fouriertransform function are peak values, the magnitude is divided by $\sqrt{2}$ to obtain the value in RMS.

```
function[i,iphase] = fouriertransform(N,samplestart,I)

% Isolate one cycle of data
k = 1;
for i = samplestart:(samplestart+N)
    y(k) = I(i);
    k = k+1;|
end



% Apply FFT
m = abs(fft(y,N));
p = angle(fft(y,N));



% Magnitude of the fundamental frequency in peak. So divide by sqrt(2) to convert
to RMS
i = (m(2)*2/N)/sqrt(2);



% Phase angle of the fundamental frequency converted from radian to degrees
iphase = p(2)*180/pi;
```

The calculation of the fault current phasor can be complicated by the presence of transients such as an exponentially decaying DC offset, which makes the fault current asymmetrical. Hence, it is imperative to choose the best cycle of fault data available in the event report to apply FFT and perform fault location analysis. It is recommended to choose a cycle in the steady-state fault portion of the event report after the transients have died out for extracting the fault phasors. For faults that start with a lower current magnitude and then increase to a higher value, the fault location window should start when the fault current transitions to the higher magnitude. This will yield the best fault location result as the fault resistance (and therefore, error) is lowest. If the fault type evolves from one type to another within the same event report, choose a window such that the waveform encompassed is of the fault type entered in Step 3.

For this purpose, after importing the current and voltage waveforms, the following code of the MATLAB script automatically plots the current in one of the faulted phases (for both terminals) against the sample number. The MATLAB script uses the fault type input entered in Step 1 to plot one of the faulted phases correctly.

```
figure3 = figure(3);
axes1 = axes('Parent',figure3,'FontSize',12);
if strcmp(Fault_type,'AG')==1 || strcmp(Fault_type,'AB')==1 || strcmp(Fault_type,'AB-G')==1
|| strcmp(Fault_type,'ABC')==1
    plot(I_A_waveG,'r','linewidth', 1);
    grid;
    if strcmp(Remote,'Yes')== 1
        hold on;
        plot(I_A_waveH,'b','linewidth', 1);
    end
    ylabel('IA (kA)','fontsize', 12);
elseif strcmp(Fault_type,'BG')==1 || strcmp(Fault_type,'BC')==1 || strcmp(Fault_type,'BC-
G')==1 || strcmp(Fault_type,'ABC')==1
    plot(I_B_waveG,'r','linewidth', 1);
    grid;
    if strcmp(Remote,'Yes')== 1
        hold on;
        plot(I_B_waveH,'b','linewidth', 1);
    end
    ylabel('IB (kA)','fontsize', 12);
else
    plot(I_C_waveG,'r','linewidth', 1);
    grid;
    if strcmp(Remote,'Yes')== 1
        hold on;
        plot(I_C_waveH,'b','linewidth', 1);
    end
    ylabel('IC (kA)','fontsize', 12);
end
xlim([1 2500]);
ylim([-10 10]);
xlabel('Number of samples','fontsize', 12);
if strcmp(Remote,'Yes')== 1
    legend('Terminal G', 'Terminal H','location','Northwest')
else
    legend('Terminal G','location','Northwest')
end
title('Faulted Phase Current vs. Sample Number','fontsize', 12)
```

Figure A.2 is useful for identifying the sample numbers denoted by *fault_inst_G* and *fault_inst_H* for extracting the fault phasors from the event report. These sample points are the starting points of the one-cycle long window to calculate the voltage and current phasors at terminals G and H during the fault using FFT. Similarly, the sample numbers denoted by *prefault_inst_G* and *prefault_inst_H* are the starting points of the one-cycle window to calculate the prefault voltage and current phasors at both terminals.

DFRs at the two ends of the line can be triggered at different time instants as in this case or can be of different sampling frequency. As a result, *fault_inst_G* and *fault_inst_H* need not be the same. In this example, it can be observed from Figure A.2 that the fault current at terminal G is asymmetrical due to the presence of a significant DC offset. To avoid any error in phasor computation, we wait for the DC offset to decay and then perform FFT on

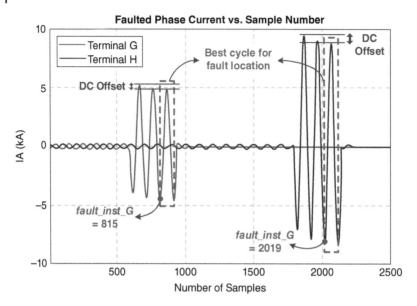

Figure A.2 Plot of currents recorded at terminals G and H in one of the faulted phases versus the sample number.

the third cycle after fault inception. Therefore, sample number 815 is a good sample point to extract the fault phasors recorded at terminal G. Similarly, for terminal H, sample number 2019 is a good sample point to extract the fault phasors after the DC offset has decayed.

The prefault instant (*prefault_inst_G*) can take any sample number before the onset of the fault. However, to avoid any phase angle mismatch between the prefault and fault phasors, the difference in the number of samples between the fault instant and prefault instant at which the respective phasors are calculated must be an integer multiple of the sampling rate, as shown in (A.3).

$$\frac{fault\_inst\_G - prefault\_inst\_G}{N\_G} = \text{Integer value.} \tag{A.3}$$

This condition is checked by the code snippet shown below. If the criterion is not satisfied, the user is asked to reenter the values of the fault (*fault_inst_G*) and prefault instants (*prefault_inst_G*). In this example, sample number 315 and 1519 were chosen as the prefault instants for the event reports recorded at terminals G and H, respectively. The difference in the number of samples between the fault instant and prefault instant is 500 for both event reports, which results in an integer multiple of the sampling rate (which is 100 samples per cycle for the recorders at both terminals).

```
integerTest = ~mod(((fault_inst_G-prefault_inst_G)/N_G),1);
if integerTest==0
    disp('The difference between prefault and fault sample number must be an exact
    multiple of the sampling rate. Enter again.');
    fprintf('\n')
end
```

The user can input the above-selected sample numbers for the prefault and fault instants in the MATLAB prompt generated by the following code.

```
prompt = 'What is the prefault instant/sampling number (Terminal G)? \n ';
prefault_inst_G = input(prompt);
disp(' ');

prompt = 'What is the fault instant/sampling number (Terminal G)? \n ';
fault_inst_G = input(prompt);
disp(' ');
```

The master script invokes the fouriertransform.m function file to calculate the phasors at the instants chosen by the user. For example, the following lines of code in the master script calculates the fault current phasors at terminal G.

```
% Phase A
[I_A_Mag_G,I_A_Phase_G] = fouriertransform(N_G,fault_inst_G,I_A_waveG);
I_A_G = I_A_Mag_G*(cosd(I_A_Phase_G) + sind(I_A_Phase_G)*1i);

% Phase B
[I_B_Mag_G,I_B_Phase_G] = fouriertransform(N_G,fault_inst_G,I_B_waveG);
I_B_G = I_B_Mag_G*(cosd(I_B_Phase_G) + sind(I_B_Phase_G)*1i);

% Phase C
[I_C_Mag_G,I_C_Phase_G] = fouriertransform(N_G,fault_inst_G,I_C_waveG);
I_C_G = I_C_Mag_G*(cosd(I_C_Phase_G) + sind(I_C_Phase_G)*1i);
```

Similarly, the master script calls the fouriertransform.m function file to calculate prefault current phasors as it is required as an input by several impedance-based fault location algorithms such as the Takagi, Eriksson et al., and Novosel et al. methods. Then, the script proceeds to compute the fault and prefault voltage phasors for both event reports.

Step 6: Apply fault location algorithms

The next section of the script gathers the input variables required for applying the fault location algorithms based on the fault type and line configuration. The two-ended algorithms implemented in this fault location suite use sequence components of the voltage and current phasors. For this purpose, a sequence transformation matrix (E) for transforming the voltage and current phasor quantities to the sequence domain is defined. This sequence transformation matrix depends on whether the power system has an ABC or ACB phase rotation. Based on the phase rotation input provided by the user in Step 3, the sequence components are calculated. In this example, the system has a positive-sequence (ABC) phase rotation.

```
a = cosd(120) + sind(120)*1i;
if strcmp(Rotation,'ABC')==1
    E = [1 1 1; 1 a^2 a; 1 a a^2];
else
    E = [1 1 1; 1 a a^2; 1 a^2 a];
end
```

Recall the importance of choosing the correct phase as the reference for calculating the sequence components from Chapter 2. For a single line-to-ground fault, the phase experiencing the fault must be used as a reference. For example, the following code shows the calculation of symmetrical components at terminal G using phase A as the base for an SLG fault on phase A.

```
case 'AG'
    V_G_012 = E\[V_A_G;V_B_G;V_C_G];
    I_G_012 = E\[I_A_G;I_B_G;I_C_G];
```

For line-to-line or double line-to-ground faults, the healthy phase must be used as the reference when calculating the zero-, positive-, and negative-sequence components. For example, phase C is used as the reference for transforming the voltage and current phasors at terminal G to the sequence domain during a line-to-line fault between phases A and B, as shown below.

```
case {'AB','AB-G'}
    I_G_012 = E\[I_C_G;I_A_G;I_B_G];
    V_G_012 = E\[V_C_G;V_A_G;V_B_G];
```

For a three-phase fault, any one of the three phases can be used as the reference when calculating the symmetrical components of the voltage and current phasors. This code uses phase A as the reference for three-phase faults.

```
case {'ABC'}
    I_G_012 = E\[I_A_G;I_B_G;I_C_G];
    V_G_012 = E\[V_A_G;V_B_G;V_C_G];
```

The positive-, negative-, and zero-sequence source impedance behind terminal G are calculated from the event report as shown below.

```
if isempty(Z_G_1)==1
    Z_G_1 = (-1)*((V_G_1-V_G_prefault_ph)/(I_G_1-I_G_prefault_ph));
end
```

```
if isempty(Z_G_2)==1
    Z_G_2 = (-1)*(V_G_2/I_G_2);
end
```

```
if isempty(Z_G_0)==1
    Z_G_0 = (-1)*(V_G_0/I_G_0);
end
```

Similar equations are used to compute the source impedance parameters behind terminal H as well if the user has the event report recorded at terminal H. If the event report is not available, the script gives the user the option to enter the remote source impedance parameters manually at this point.

Based on the fault type, the MATLAB script uses the switch function to define the required input parameters such as current and voltage phasors and executes the applicable impedance-based fault location algorithms.

```
switch Fault_type
    case 'AG'
        Input parameters
        ...
        Fault location algorithms

        |..|
        |..|

    case 'CG'
        Input parameters
        ...
        Fault location algorithms

        |..|
        |..|

    case {'AB','AB-G'}
        Input parameters
        ...
        Fault location algorithms

        |..|
        |..|

    case 'ABC'
        Input parameters
        ...
        Fault location algorithms
end
```

The last part of this step explains how the fault location algorithms are implemented in the MATLAB code.

1) One-Ended Fault Location Methods
   (a) Simple Reactance Method
       To apply this method, the input voltage and current phasors (V_G and I_G) are computed based on the fault type. For a single line-to-ground fault, they are calculated as

```
% Residual compensation for single line-to-ground faults
k = (Z_0-Z_1)/Z_1;

% Fault voltage and current phasors during a single line-to-ground fault
V_G = V_ph;
I_G = I_ph + k*I_G_0;
```

For all other fault types, they are calculated as

```
% Fault and voltage current phasors during a LL, LL-G, or a LLL fault
V_G = V_ph1 - V_ph2;
I_G = I_ph1 - I_ph2;
```

The simple reactance method computes the distance to the fault (m) using (3.10) as

```
% Estimate from the Simple Reactance Method
m = imag(V_G/I_G)/imag(Z_1); % pu
D_SR = m*LL; % unit of line length
```

(b) Takagi Method

The Takagi method computes the distance to the fault using (3.15) as

```
% Estimate from the Takagi Method
m = (imag(V_G*conj(del_I_G))/imag(Z_1*I_G*conj(del_I_G))); % pu
D_Takagi = m*LL; % unit of line length
```

where ΔI_G current phasor is calculated based on the fault type as defined in Table 3.2. For single line-to-ground faults, ΔI_G is calculated as

```
% 'Pure fault' current
del_I_G = I_ph - I_prefault_ph;
```

For all other fault types, it is calculated as

```
% 'Pure fault' current
del_I_G = (I_ph1 - I_prefault_ph1) - (I_ph2 - I_prefault_ph2);
```

(c) Modified Takagi Method

For single line-to-ground faults, the modified Takagi method computes the distance to fault as

```
% Estimate from the Modified Takagi Method
m = (imag(V_G*conj(I_G_2))/imag(Z_1*I_G*conj(I_G_2))); % pu
D_ModTakagi = m*LL; % unit of line length
```

For all other fault types, it is calculated as

```
% Estimate from the Modified Takagi Method
if strcmp(Fault_type,'ABC')==1
    m = (imag(V_G*conj(I_G))/imag(Z_1*I_G*conj(I_G))); % pu
    D_ModTakagi = m*LL; % unit of line length
else
    m = (imag(V_G*conj(1i*I_G_2))/imag(Z_1*I_G*conj(1i*I_G_2))); % pu
    D_ModTakagi = m*LL; % unit of line length
end
```

(d) Eriksson et al. Method (Current Distribution Factor Method)

The Eriksson method calculates the distance to fault (m) as a solution to the quadratic equation discussed in Section 3.1.4.

```
% Solve the quadratic equation
m(1) = (-B+sqrt(B^2-(4*A*C)))/(2*A);
m(2) = (-B-sqrt(B^2-(4*A*C)))/(2*A);
```

The constants of the quadratic equation are defined as

```
% Calculate the constants k1, k2, and k3
k1 = 1 + (Z_H_1/Z_1) + (V_G/(Z_1*I_G));
k2 = (V_G/(Z_1*I_G))*(1+(Z_H_1/Z_1));
k3 = ((del_I_G)/(Z_1*I_G))*(1+((Z_H_1+Z_G_1)/Z_1));
a = real(k1);
b = imag(k1);
c = real(k2);
d = imag(k2);
e = real(k3);
f = imag(k3);
A = 1;
B = -a+(e*b)/f;
C = c-(e*d)/f;
```

As m can take two possible values, the value of m that lies between 0 and 1 per unit is chosen as the fault location estimate.

```
% m must be less than the total line length. So choose the value of m that lies between 0
and 1 per unit.
for element = 1:2
    if m(element)>0 && m(element)<1
        D_Eriksson = m(element)*LL;
    end
end
```

(e) Novosel et al. Method

As discussed earlier in Step 4, the Novosel et al. method is applicable for locating faults on radial transmission lines. The first step is to estimate the load impedance using the prefault current and voltage phasors.

```
% Load Impedance
Z_Load = (V_prefault_ph/I_prefault_ph)-Z_1;
```

The next step is to solve for the distance to fault estimate from the quadratic equation discussed in Section 5.1.

```
% Solve the quadratic equation
m(1) = (-B+sqrt(B^2-(4*A*C)))/(2*A);
m(2) = (-B-sqrt(B^2-(4*A*C)))/(2*A);
```

where the constants are defined as

```
% Calculate the constants k1, k2, and k3
k1 = 1 + (Z_Load/Z_1) + (V_G/(Z_1*I_G));
k2 = (V_G/(Z_1*I_G))*(1+(Z_Load/Z_1));
k3 = ((del_I_G)/(Z_1*I_G))*(1+((Z_Load+Z_G_1)/Z_1));
a = real(k1);
b = imag(k1);
c = real(k2);
d = imag(k2);
e = real(k3);
f = imag(k3);
A = 1;
B = -a+(e*b)/f;
C = c-(e*d)/f;
```

Finally, the value of *m* that lies between 0 and 1 per unit is chosen as the final location estimate.

```
% m must be less than the total line length. So choose the value of m that lies between 0
and 1 per unit.
for element = 1:2
    if m(element)>0 && m(element)<1
        D_Novosel = m(element)*LL;
    end
end
```

2) Two-Ended Fault Location Methods
  (a) Synchronized Two-Ended Method
    The synchronized two-ended method calculates the distance to the fault using (3.27) and (3.28) as

```
m = (V_G_i-V_H_i+(Z_i*I_H_i))/((I_G_i+I_H_i)*Z_i);  % pu

D_Sync2ended = m*LL;  % in units of line length
```

where *i* refers to the ith symmetrical component. As discussed in Section 3.2.1, the negative-sequence components ($i = 2$) are used for unbalanced faults and the positive-sequence components ($i = 1$) are used for symmetrical three-phase faults.
  (b) Unsynchronized Two-Ended Method
    The unsynchronized two-ended method estimates the distance to the fault (*m*) as the solution to a quadratic equation discussed in Section 3.2.2. The value of *m* is calculated using (3.33) as shown below.

```
% Solve for m as a quadratic equation
m(1) = (-B+sqrt(B^2-(4*A*C)))/(2*A);
m(2) = (-B-sqrt(B^2-(4*A*C)))/(2*A);
```

where the constants are defined as

```
% Defining the constants
A = abs(Z_i*I_G_i)^2-abs(Z_i*I_H_i)^2;
B = (-2)*real((V_G_i*conj(Z_i*I_G_i))+((V_H_i-(Z_i*I_H_i))*conj(Z_i*I_H_i)));
C = abs(V_G_i)^2-abs(V_H_i-(Z_i*I_H_i))^2;
```

Here again, i refers to the ith symmetrical component. The negative-sequence components ($i = 2$) are used for unbalanced faults and the positive-sequence components ($i = 1$) are used for symmetrical three-phase faults. Finally, the value of m that lies between 0 and 1 per unit is chosen as the fault location estimate.

```
% Choose the value of m that lies between 0 and 1 pu
for element = 1:2
    if m(element)>0 && m(element)<1
        D_Unsync2ended = m(element)*LL;
    end
end
```

(c) Unsynchronized Negative-Sequence Two-Ended Method

The unsynchronized negative-sequence two-ended method uses only negative-sequence current phasors and source impedance parameters to determine the fault location (m). This method estimates the distance to the fault as the solution to a quadratic equation discussed in Section 3.2.3.

```
% Solving the quadratic equation
m(1) = (-B+sqrt(B^2-(4*A*C)))/(2*A);
m(2) = (-B-sqrt(B^2-(4*A*C)))/(2*A);
```

where the constants are defined as

```
% Define the constants
a = real(I_G_2*Z_G_2);
b = imag(I_G_2*Z_G_2);
c = real(Z_2*I_G_2);
d = imag(Z_2*I_G_2);
e = real(Z_H_2+Z_2);
f = imag(Z_H_2+Z_2);
g = real(Z_2);
h = imag(Z_2);

% Coefficients of the quadratic equation
A = (abs(I_H_2))^2*(g^2+h^2)-(c^2+d^2);
B = (-2)*(abs(I_H_2))^2*((e*g)+(f*h))-2*((a*c)+(b*d));
C = (abs(I_H_2))^2*(e^2+f^2)-(a^2+b^2);
```

The value of m between 0 and 1 per unit is then chosen as the distance to the fault estimate. As this method uses only negative-sequence components, it is not applicable for locating three-phase faults.

When the input information is incorrect or the calculated fault location estimate does not lie between 0 and 1 pu of the line length for some algorithms, the MATLAB software may halt and show an error. In such cases, check the input information and run the master script again. If the error persists, observe the results from methods such as simple reactance or Takagi method in the MATLAB workspace. If the fault location estimate from these methods were outside the line length, it would likely be the reason for the error message.

Step 7: View the fault location estimates from the fault location algorithms

The following code displays the fault location estimates calculated using all the methods discussed in this section in the MATLAB command window.

```
disp('******************************************************************************');
disp('Event Report:');
disp(' ');
fprintf('Fault type: %s\n',Fault_type);
disp(' ');
disp('******************************************************************************');
disp('One-ended Fault Location Results (Distance from Terminal G):');
disp(' ');
fprintf('Simple reactance: %0.2f %s \n', D_SR, Line_Length_units);
disp(' ');
fprintf('Takagi: %0.2f %s \n',D_Takagi, Line_Length_units);
disp(' ');
if isnumeric(D_ModTakagi)==1
    fprintf('Modified Takagi: %0.2f %s \n', D_ModTakagi, Line_Length_units)
else
    fprintf('Modified Takagi: %s \n', D_ModTakagi)
end
disp(' ');
if isnumeric(D_Eriksson)==1
    fprintf('Eriksson et al.: %0.2f %s \n', D_Eriksson, Line_Length_units)
else
    fprintf('Eriksson et al.: %s \n', D_Eriksson)
end
disp(' ');
if isnumeric(D_Novosel)==1
    fprintf('Novosel et al.: %0.2f %s \n', D_Novosel, Line_Length_units)
else
    fprintf('Novosel et al.: %s \n', D_Novosel)
end
disp(' ');
disp('******************************************************************************');
disp('Two-ended Fault Location Results (Distance from Terminal G):');
disp(' ');
if isnumeric(D_Sync2ended)==1
    fprintf('Synchronized Two-ended: %0.2f %s \n', D_Sync2ended, Line_Length_units)
else
    fprintf('Synchronized Two-ended: %s \n', D_Sync2ended)
end
disp(' ');
if isnumeric(D_Unsync2ended)==1
    fprintf('Unsynchronized Two-ended: %0.2f %s \n', D_Unsync2ended, Line_Length_units)
else
    fprintf('Unsynchronized Two-ended: %s \n', D_Unsync2ended)
end
disp(' ');
if isnumeric(D_UnsyncCurrentOnly)==1
    fprintf('Unsynchronized Negative-Sequence Two-ended: %0.2f %s \n', D_UnsyncCurrentOnly,
Line_Length_units)
else
    fprintf('Unsynchronized Negative-Sequence Two-ended: %s \n', D_UnsyncCurrentOnly)
end
disp(' ');
disp('******************************************************************************');
```

```
********************************************************************
Event Report:

Fault type: AG

********************************************************************
One-ended Fault Location Results (Distance from Terminal G):

Simple reactance: 14.80 miles

Takagi: 14.79 miles

Modified Takagi: 14.79 miles

Eriksson et al.: 14.80 miles

Novosel et al.: Not Applicable

********************************************************************
Two-ended Fault Location Results (Distance from Terminal G):

Synchronized Two-ended: Not Applicable

Unsynchronized Two-ended: 14.77 miles

Unsynchronized Negative-Sequence Two-ended: 14.77 miles

********************************************************************
```

Figure A.3 MATLAB command window displaying the calculated fault location estimates.

Figure A.3 shows the fault location estimates calculated using the fault location suite for this example scenario. The Novosel et al. method could not be applied because the transmission line in this example was not radial. The synchronized two-ended methods could not be applied because the measurement data from both ends of the line were not synchronized. Recollect that the fault was present at 14.90 miles from terminal G. Hence, both one- and two-ended algorithms were successful in estimating the location of the fault accurately.

References

1 (2014) IEEE standard for seismic qualification testing of protective relays and auxiliaries for nuclear facilities. *IEEE Std C37.98-2013 (Revision of IEEE Std C37.98-1987)*, pp. 1–35.

2 Behrendt, K. (2011) Detecting high-side blown fuse/open-phase conditions using an SEL-651R recloser control, Schweitzer Engineering Laboratories, *AG2011-19*, vol. III.

3 Edmund O. Schweitzer, I. and Kumm, J.J. (1997) Detecting high-side fuse operations using an SEL-351 relay, Schweitzer Engineering Laboratories, *AG97-11*, vol. III.

4 Rusicior, M., Young, J., and Burfield, A. (2016) Open-phase detection in the SEL-751 and SEL-751A feeder protection relays, Schweitzer Engineering Laboratories, *AG2016-07*, vol. III.

5 (1992) Single phase tripping and auto reclosing of transmission lines-IEEE committee report. *IEEE Transactions on Power Delivery*, vol. 7, no. 1, pp. 182–192.

6 Kojovic, L.A. and Williams, C.W. (2000) Sub-cycle detection of incipient cable splice faults to prevent cable damage, in *2000 Power Engineering Society Summer Meeting (Cat. No.00CH37134)*, vol. 2, pp. 1175–1180.

7 IEEE Power Systems Relaying Committee (1984) Automatic reclosing of transmission lines. *IEEE Transactions on Power Apparatus and Systems*, **PAS-103** (2), pp. 234–245.

8 (1993) IEEE guide for liquid-immersed transformers through-fault-current duration. *IEEE Std C57.109-1993*, pp. 1–125.

9 Buff, J. and Zimmerman, K. (2007) Application of existing technologies to reduce arc-flash hazards, in *2007 60th Annual Conference for Protective Relay Engineers*, pp. 218–225.

10 Benmouyal, G., Hou, D., and Tziouvaras, D. (2004) Zero-setting power-swing blocking protection, in *Proc. 31st Annual Western Protective Relay Conference*, pp. 1–29.

11 Perumalsamy, M., Kapse, S., and Barikar, C. (2019) Extremely fast tripping for an arc-flash event: A field case analysis, in *46th Annual Western Protective Relay Conference*, pp. 1–12.

12 North American Electric Reliability Corporation (2020) An assessment of 2019 bulk power system performance, in *2020 State of Reliability*.

Fault Location on Transmission and Distribution Lines: Principles and Applications, First Edition.
Swagata Das, Surya Santoso, and Sundaravaradan N. Ananthan.
© 2022 John Wiley & Sons Ltd. Published 2022 by John Wiley & Sons Ltd.
Companion website: www.wiley.com/go/das/faultlocation

13 Cummins, K.L. and Murphy, M.J. (2009) An overview of lightning locating systems: History, techniques, and data uses, with an in-depth look at the U.S. NLDN. *IEEE Transactions on Electromagnetic Compatibility*, vol. 51, no. 3, pp. 499–518.

14 Boecker, M., Corpuz, G., Hargrave, G., Das, S., Fischer, N., and Skendzic, V. (2018) Line current differential relay response to a direct lightning strike on a phase conductor, in *2018 71st Annual Conference for Protective Relay Engineers*, pp. 1–12.

15 CN Utility Consulting, L. (2004) Utility vegetation management final report. *Federal Energy Regulatory Commission*.

16 Fieldstadt, E. (2016) Texas hot air balloon likely hit power lines before crash: Feds. *NBC News*.

17 Smith, R. (2014) Assault on california power station raises alarm on potential for terrorism. *Wall Street Journal*.

18 Texas Reliability Entity, Inc. (2016) 2015 assessment of reliability performance of the Electric Reliability Council of Texas, Inc. (ERCOT) region.

19 (Jan 2017) Transforming the nation's electricity system: The second installment of the QER. *US Department of Energy*.

20 Nicholson, L.C. (1907) Location of broken insulators and other transmission line troubles. *Proceedings of the American Institute of Electrical Engineers*, vol 26, no. 5, pp. 723–733.

21 Holbeck, J.I. (1944) A simple method for locating ground faults. *Electrical Engineering*, vol 63, no. 3, pp. 89–92.

22 Spaulding, L.R. and Diemond, C.G. (1950) Fault locator for high-voltage lines. *Electrical Engineering*, vol. 69, no. 2, pp. 134–134.

23 Saha, M.M., Izykowski, J.J., and Rosolowski, E. (2010) *Fault Location on Power Networks*, Springer-Verlag, New York, USA, 1st edn.

24 Saravanan, N. and Rathinam, A. (2012) A comparitive study on ANN based fault location and classification technique for double circuit transmission line, in *2012 Fourth International Conference on Computational Intelligence and Communication Networks*, pp. 824–830.

25 Babayomi, O., Oluseyi, P., Keku, G., and Ofodile, N.A. (2017) Neuro-fuzzy based fault detection identification and location in a distribution network, in *2017 IEEE PES PowerAfrica*, pp. 164–168.

26 Mooney, J. (1996) Microprocessor-based transmission line relay applications. *American Public Power Associations Engineering and Operations Workshop*.

27 Kirby, R.D. and Schwartz, R. (2006) Microprocessor-based protective relays deliver more information and superior reliability with lower maintenance costs, in *2006 IEEE Industrial and Commercial Power Systems Technical Conference - Conference Record*, pp. 1–7.

28 Working Group I-01 (2009) Understanding microprocessor-based technology applied to protective relaying. *Power System Relaying Committee*.

29 Costello, D. (2008) Lessons learned analyzing transmission faults, in *2008 61st Annual Conference for Protective Relay Engineers*, pp. 410–422.

30 Perez, J. (2010) A guide to digital fault recording event analysis, in *2010 63rd Annual Conference for Protective Relay Engineers*, pp. 1–17.

31 Ametek Power Instruments (2006) TR-100+ multi-function recorder operation manual, in *Publication 1079-305, Rev B*.

32 Qualitrol BEN 6000 extreme digital fault recorder bulletin.

33 Schweitzer Engineering Laboratories, Washington, Pullman *SEL-T400L Instruction Manual*.

34 Qualitrol TWS FL-8 traveling wave fault locator user manual, in *Document ID: 40-08591-01*.

35 Schweitzer Engineering Laboratories, Washington, Pullman *SEL-5045 Instruction Manual*.

36 (2015) IEEE guide for determining fault location on AC transmission and distribution lines. *IEEE Std C37.114-2014 (Revision of IEEE Std C37.114-2004)*, pp. 1–76.

37 Fortescue, C. (1918) Method of symmetrical co-ordinates applied to the solution of polyphase networks, in *34th Annual Convention of the American Institute of Electrical Engineers*.

38 J.L. Blackburn and T.J. Domin (2007) *Protective Relaying Principles and Applications*, CRC Press, Boca Raton, FL, USA, 3rd edn.

39 Stevenson, Jr., W.D. (1982) *Elements of Power System Analysis*, McGraw Hill, 4th edn.

40 Working Group D6 (2014) Ac transmission line model parameter validation. *A report to the Line Protection Subcommittee of the Power System Relay Committee of the IEEE Power & Energy Society*.

41 Amberg, A. and Rangel, A. (2013) Tutorial on symmetrical components part 1: Examples, Schweitzer Engineering Laboratories.

42 Amberg, A. and Rangel, A. (2020) Tutorial on symmetrical components part 2: Answer key, Schweitzer Engineering Laboratories.

43 Schweitzer, III, E.O. and Zocholl, S. (2004) Introduction to symmetrical components, in *58th Annual Georgia Tech Protective Relaying Conference*, pp. 1–16.

44 Glover, J., Sarma, M., and Overbye, T. (2012) *Power System Analysis and Design*, Cengage Learning, 5th edn.

45 Kersting, W.H. (2012) *Distribution System Modeling and Analysis*, CRC Press, Boca Raton, FL, USA, 3rd edn.

46 Horton, R., Sunderman, W., Arritt, R., and Dugan, R. (2011) Effect of line modeling methods on neutral-to-earth voltage analysis of multi-grounded distribution feeders, in *2011 IEEE/PES Power Systems Conference and Exposition. (PSCE)*, pp. 1–6.

47 Zocholl, S. (2004) Symmetrical components: Line transposition, in *Document 6606 Schweitzer Engineering Laboratories, Inc.*, pp. 1–6.

48 Dierks, A., Troskie, H., and Kruger, M. (2005) Accurate calculation and physical measurement of trasmission line parameters to improve impedance relay performance, in *2005 IEEE Power Engineering Society Inaugural Conference and Exposition in Africa*, pp. 455–461.

49 K. Thomas (2014) Comparison of methods to determine transmission line impedance, in *Georgia Tech Fault and Disturbance Analysis Conference*.

50 Takagi, T., Yamakoshi, Y., Yamaura, M., Kondow, R., and Matsushima, T. (1982) Development of a new type fault locator using the one-terminal voltage and current data. *IEEE Trans. Power App. Syst.*, **PAS-101** (8), 2892–2898.

51 Zimmerman, K. and Costello, D. (2005) Impedance-based fault location experience, in *Proc. 58th Annu. Conf. Protect. Relay Eng.*, pp. 211–226.

52 Živanović, R. (2008) Evaluation of transmission line fault-locating techniques using variance-based sensitivity measures, in *Proc. 16th Power Syst. Comput. Conf.*, pp. 1–6.

53 Eriksson, L., Saha, M.M., and Rockefeller, G.D. (1985) An accurate fault locator with compensation for apparent reactance in the fault resistance resulting from remote-end infeed. *IEEE Trans. Power App. Syst.*, vol. PAS-104, no. 2, pp. 423–436.

54 Schweitzer III, E.O. (1990) A review of impedance-based fault locating experience, in *Proc. 14th Annu. Iowa-Nebraska Syst. Protect. Seminar*.

55 Amberg, A., Rangel, A., and Smelich, G. (2012) Validating transmission line impedances using known event data, in *Proc. 65th Annu. Conf. Protect. Relay Eng.*, pp. 269–280.

56 Schweitzer III, E.O. (1983) Evaluation and development of transmission line fault-locating techniques which use sinusoidal steady-state information. *Comput. Elect. Eng.*, vol. 10, no. 4, pp. 269–278.

57 Tziouvaras, D.A., Roberts, J.B., and Benmouyal, G. (2001) New multi-ended fault location design for two- or three-terminal lines, in *Proc. IEE 7th Int. Conf. Develop. Power Syst. Protect.*, pp. 395–398.

58 Kaszenny, B., Le, B., and Fischer, N. (2012) A new multiterminal fault location algorithm embedded in line current differential relays, in *Proc. 11th Int. Conf. Develop. Power Syst. Protect.*

59 North American Electric Reliability Council (2006), The complexity of protecting three-terminal transmission lines.

60 L.V. Bewley *Traveling Waves on Transmission Systems*, Dover Publications, New York, 2nd edn.

61 Kasztenny, B. (2019) Improving line crew dispatch accuracy when using traveling-wave fault locators, in *46th Annual Western Protective Relay Conference*, pp. 1–13.

62 Schweitzer, E.O., Guzmán, A., Mynam, M.V., Skendzic, V., Kasztenny, B., and Marx, S. (2014) A new traveling wave fault locating algorithm for line current differential relays, in *12th International Conference on Developments in Power System Protection*, pp. 1–6.

63 Kasztenny, B., Guzmán, A., Mynam, M.V., and Joshi, T. (2018) Locating faults before the breaker opens adaptive autoreclosing based on the location of the fault, in *2018 71st Annual Conference for Protective Relay Engineers*, pp. 1–15.

64 E.O. Schweitzer, Guzmán, A., M.V. Mynam, V. Skendzic, B.Kasztenny, and S. Marx (2014) Locating faults by the traveling waves they launch, in *2014 67th Annual Conference for Protective Relay Engineers*, pp. 95–110.

65 J. Trejo and D. Cole (2017) A comparative study on cost savings offered by traveling wave system fault locators over traditional distance relays for long distance transmission lines, in *Georgia Tech Fault and Disturbance Analysis Conference*.

66 Fischer, N., Skendzic, V., Moxley, R., and Needs, J. (2012) Protective relay traveling wave fault location, in *11th International Conference on Developments in Power System Protection*.

67 Guzmán, A., Kasztenny, B., Tong, Y., and Mynam, M.V. (2018) Accurate and economical traveling-wave fault locating without communications, in *2018 71st Annual Conference for Protective Relay Engineers (CPRE)*, pp. 1–18.

68 E.O. Schweitzer, A. Guzmán, M. Mynam, V. Skendzic, and B. Kasztenny (2016) Accurate single-end fault location and line-length estimation using traveling waves, in *13th International Conference on Developments in Power System Protection, March 2016*, pp. 1–6.

69 D. Cole, M.Diamond, and A.Kulshrestha (2012) Fault location and system restoration, in *PAC World*, vol. September 2012 issue.

70 Polikoff, A. (2015) Using traveling wave fault location in the SEL-411L-1 relay, in *AG2015-20*, vol. 1, pp. 1–10.

71 Short, T.A. (2004) *Electric Power Distribution Handbook*, CRC Press, Boca Raton, FL, USA.

72 (EPRI, Palo Alto, CA: 2006. 1012438) *Distribution fault location: field data and analysis.*

73 Schweitzer III, E.O. and Hou, D. (1993) Filtering requirements for distance relays, in *Proc. American Power Conf.*, pp. 296–301.

74 Hargrave, A., Thompson, M.J., and Heilman, B. (2018) Beyond the knee point: A practical guide to CT saturation, in *2018 71st Annual Conference for Protective Relay Engineers*, pp. 1–23.

75 Benmouyal, G., Roberts, J., and Zocholl, S. (1996) Selecting CTs to optimize relay performance, in *23rd Annual Western Protective Relay Conference*, pp. 1–9.

76 Gray, S., Haas, D., and McDaniel, R. (2018) CCVT failures and their effects on distance relays, in *2018 71st Annual Conference for Protective Relay Engineers*, pp. 1–13.

77 Costello, D. and Zimmerman, K. (2012) CVT transients revisited - distance, directional overcurrent, and communications-assisted tripping concerns, in *2012 65th Annual Conference for Protective Relay Engineers*, pp. 73–84.

78 Fehr, R.E. (2004), Sequence impedances of transmission lines.

79 Markiewicz, H. and Klajn, A. (2003) Earthing systems - fundamentals of calculation and design, in *Power Quality Application Guide*, Copper Development Association, Hertfordshire, U.K.

80 Sharma, D., Kathe, A., Joshi, T., and Kanagasabai, T. (2019) Application of ultra-high-speed protection and traveling-wave fault locating on a hybrid line, in *2019 46th Annual Western Protective Relay Conference*, pp. 1–12.

81 Tziouvaras, D. (2006) Protection of high-voltage ac cables, in *59th Annual Conference for Protective Relay Engineers, 2006.*, pp. 48–61.

82 Costello, D. and Zimmerman, K. (2010) Determining the faulted phase, in *2010 63rd Annual Conference for Protective Relay Engineers*, pp. 1–20.

83 Kasztenny, B., Campbell, B., and Mazereeuw, J. (2010) Phase selection for single-pole tripping weak infeed conditions and cross-country faults, in *2000 27th Annual Western Protective Relay Conference*, pp. 1–19.

84 Tziouvaras, D.A., Altuve, H.J., and Calero, F. (2014) Protecting mutually coupled transmission lines: Challenges and solutions, in *2014 67th Annual Conference for Protective Relay Engineers*, pp. 30–49.

85 Holt, C. and Thompson, M.J. (2016) Practical considerations when protecting mutually coupled lines, in *2016 69th Annual Conference for Protective Relay Engineers*, pp. 1–16.

86 Altuve, H.J., Mooney, J.B., and Alexander, G.E. (2009) Advances in series-compensated line protection, in *2009 62nd Annual Conference for Protective Relay Engineers*, pp. 263–275.

87 Bakie, E., Westhoff, C., Fischer, N., and Bell, J. (2016) Voltage and current inversion challenges when protecting series-compensated lines - a case study, in *2016 69th Annual Conference for Protective Relay Engineers*, pp. 1–14.

88 G.E. Alexander, S.D. Rowe, J.G. Andrichak, and S.B.Wilkinson Series compensated line protection - a practical evaluation, in *GER-3736*, pp. 1–52.

89 Santoso, S. (2009) *Fundamentals of Electric Power Quality*, CreateSpace, Scotts Valley, CA, USA, Spring edn.

90 M.S. Elkateb (1978) Cable fault location techniques in E.H.V, H.V, M.V, and L.V. systems. *University of New South Wales Australia.*

91 Girgis, A.A., Fallon, C.M., and Lubkeman, D.L. (1993) A fault location technique for rural distribution feeders. *IEEE Transactions on Industry Applications*, vol. 29, no. 6, pp. 1170–1175.

92 Santoso, S., Dugan, R., Lamoree, J., and Sundaram, A. (2000) *Distance estimation technique for single line-to-ground faults in a radial distribution system*, vol. 4, pp. 2551–2555.

93 Novosel, D., Hart, D., Hu, Y., and Myllymaki, J. (1998), System for locating faults and estimating fault resistance in distribution networks with tapped loads. US Patent US5839093A, filed 31 December, 1996 and issued 17 November, 1998.

94 Gong, Y. and Guzmán, A. (2011) Distribution feeder fault location using IED and FCI information, in *2011 64th Annual Conference for Protective Relay Engineers*, pp. 168–177.

95 Das, S., Santoso, S., and Maitra, A. (2014) Effects of distributed generators on impedance-based fault location algorithms, in *Proc. IEEE PES General Meeting Conf. Expo.*, pp. 1–5.

96 (2009), IEEE standard for interconnecting distributed resources with electric power systems in IEEE Std. 1547–200.

97 Nunes, J.U.N. and Bretas, A.S. (2011) A impedance-based fault location technique for unbalanced distributed generation systems, in *Proc. IEEE Trondheim PowerTech*, Trondheim, pp. 1–7.

98 Barker, P.P. and De Mello, R.W. (2000) Determining the impact of distributed generation on power systems: Part 1 - radial distribution systems, in *Proc. IEEE Power Eng. Soc. Summer Meeting*, vol. **3**, vol. 3, pp. 1645–1656.

99 Samaan, N., Zavadil, R., Smith, J., and Conto, J. (2008) Modeling of wind power plants for short circuit analysis in the transmission network, in *Proc. IEEE/PES Transmission Distrib. Conf. Expo.*, pp. 1–7.

100 Ebrahimi, E., Ghanizadeh, A.J., Rahmatian, M., and Gharehpetian, G.B. (2012) Impact of distributed generation on fault locating methods in distribution networks, in *Proc. Int. Conf. Renewable Energies and Power Quality (ICREPQ'12)*, Spain, pp. 1–5.

101 Marvik, J.I. (2011) *Fault Localization in Medium Voltage Distribution Networks with Distributed Generation*, Ph.D. thesis, Norwegian University of Science and Technology, Norway.

102 Daqing Hou (2007) Detection of high-impedance faults in power distribution systems, in *2007 Power Systems Conference: Advanced Metering, Protection, Control, Communication, and Distributed Resources*, pp. 85–95.

103 Hou, D. (2009) High-impedance fault detection - field tests and dependability analysis, in *Proc. 36th Annual Western Protective Relay Conference*, pp. 1–10.

104 Theron, J.C.J., Pal, A., and Varghese, A. (2018) Tutorial on high impedance fault detection, in *2018 71st Annual Conference for Protective Relay Engineers*, pp. 1–23.

105 Line Protection Subcommittee Working Group D-7 Loss of ac voltage considerations for line protection.

106 Das, S., Karnik, N., and Santoso, S. (2012) Distribution fault-locating algorithms using current only. *IEEE Trans. Power Del.*, vol. 27, no. 3, pp. 1144–1153.

107 Das, S., Kulkarni, S., Karnik, N., and Santoso, S. (2011) Distribution fault location using short-circuit fault current profile approach, in *Proc. IEEE Power Energy Soc. General Meeting*, San Diego, CA, pp. 1–7.

108 Lampley, G. (2002) Fault detection and location on electrical distribution system, in *Proc. IEEE Rural Electric Power Conf.*, pp. B1–1–B1–5.

109 Baldick, R. (2009) *Applied Optimization: Formulation and Algorithms for Engineering Systems*,Cambridge University Press, Cambridge, England, 1st edn.

110 Schweitzer Engineering Laboratories, Washington, Pullman *SEL-351S Protection System Instruction Manual*.

111 North American Electric Reliability Corporation (NERC) (2013), Misoperations report.

112 Zimmerman, K. and Costello, D. (2009) Lessons learned from commissioning protective relaying systems, in *2009 62nd Annual Conference for Protective Relay Engineers*, pp. 359–381.

113 Welton, D. and Knapek, W. (2017) Important considerations in testing and commissioning digital protective relays, in *2017 IEEE Rural Electric Power Conference (REPC)*, pp. 12–16.

114 Labuschagne, C., and Fischer, N. (2006) Relay-assisted commissioning, in *2006 59th Annual Conference for Protective Relay Engineers College Station, Texas*, pp. 1–11.

115 Barrera Núñez, V., Kulkarni, S., Santoso, S., and Joaquim, M.F. (2010) Feature analysis and classification methodology for overhead distribution fault events, in *IEEE PES General Meeting*, pp. 1–8.

116 North American Electric Reliability Corporation (NERC) (2010), Lessons learned: Short circuit models (relay settings and equipment specifications).

117 Ward, D.J. (2003) Overhead distribution conductor motion due to short-circuit forces. *IEEE Transactions on Power Delivery*, vol. 18, no. 4, pp. 1534–1538.

118 Blair, J., Hataway, G., and Mattson, T. (2016) Solutions to common distribution protection challenges, in *2016 69th Annual Conference for Protective Relay Engineers*, pp. 1–10.

119 Ken Sanford, Arizona Public Service Co. and John S. Bowers, Pickwick Electric Cooperative (2013) Incipient faults: Can they be seen? *T&D World Magazine*.

120 (2013) IEEE/IEC Measuring relays and protection equipment - part 24: Common format for transient data exchange (COMTRADE) for power systems. *IEEE Std C37.111-2013*, pp. 1–73.

121 Prado-Félix, H., Serna-Reyna, V., Mynam, V., Donolo, M., and Guzmán, A. (2014) Improve Transmission Fault Location and Distance Protection Using Accurate Line Parameters, in *17th Georgia Tech Fault and Disturbance Analysis Conference*.

Index

Fault Location on Transmission and Distribution Lines: Principles and Applications, First Edition.
Swagata Das, Surya Santoso, and Sundaravaradan N. Ananthan.
© 2022 John Wiley & Sons Ltd. Published 2022 by John Wiley & Sons Ltd.
Companion website: www.wiley.com/go/das/faultlocation